Choice by Cable

The Economics of a New Era in Television

C. G. VELJANOVSKI
Research Officer in Law and Economics,
Centre for Socio-Legal Studies,
University of Oxford

and

W. D. BISHOP
Lecturer in Law,
London School of Economics and Political Science

TRANSATLANTIC ARTS, INC.
Sole Distributor For North America
P. O. Box 6086
ALBUQUERQUE, NM 87197 U.S.A.

Published by

THE INSTITUTE OF ECONOMIC AFFAIRS

1983

384.5556
V43c

First published in February 1983

by

The Institute of Economic Affairs
2 Lord North Street, Westminster
London SW1P 3LB

© THE INSTITUTE OF ECONOMIC AFFAIRS 1983

All rights reserved

ISSN 0073-2818

ISBN 0-255 36159-9

Printed in England by

GORON PRO-PRINT CO LTD

6 Marlborough Road, Churchill Industrial Estate, Lancing, W. Sussex

Text set in 'Monotype' Baskerville

CONTENTS

UNIVERSITY LIBRARIES
CARNEGIE-MELLON UNIVERSITY
PITTSBURGH, PENNSYLVANIA 15213

PREFACE

The *Hobart Papers* are intended to contribute a stream of authoritative, independent and lucid analyses to the understanding and application of economics to private and government activity. The characteristic theme has been the optimum use of scarce resources and the extent to which it can best be achieved in markets within an appropriate framework of law and institutions or, where markets cannot work, in other ways. Since in the real world the alternative to the market is the state, and both are imperfect, the choice between them effectively turns on a judgement of the comparative consequences of 'market failure' and 'government failure'.

One sector of economic activity where it has for long been believed in Britain that the market cannot work is broadcasting. This belief has been buttressed by a succession of official inquiries into the structure of the broadcasting industry over the past 60 years—in more recent times by Lord Pilkington's Committee in 1962 and Lord Annan's Committee in 1977. In January 1962, the Institute published as Hobart Paper 15 *TV: From Monopoly to Competition,* by Wilfred Altman, Denis Thomas and David Sawers, which argued the unfashionable case for more choice in television through more competition and more market-sensitive financing. That *Paper* went into a second edition as *TV: From Monopoly to Competition—and Back?* after Pilkington recommended a quite differently-coloured package of government controls with a maximum of monopoly and a minimum of competition. Six years later, in Hobart Paper 43, *Paying for TV?,* Sir Sydney Caine analysed the financial structure of British television and put forward a number of proposals to create a more direct financial link between those who supply television services and those who consume them. He advocated as the 'ideal' system one in which all television services, whether provided by the BBC or by commercial contractors, were supplied through pay-television receivers and financed by a combination of payment by viewers on the basis of time spent in viewing and the sale of advertising time, the TV licence fee having been abolished.

Broadcasting policy in Britain has been dominated by the assumption that broadcasting is a 'public service' and should be provided by a public authority specifically mandated to serve the public interest—or, after the advent of commercial television in 1956, by private companies operating within precise guidelines and under strict public control. That assumption derived firstly from the technology of broadcasting—the limited range of the electro-magnetic spectrum was claimed to necessitate government regulation to allocate frequencies and prevent interference—and secondly from fear of what the Annan Committee called 'the potential power [of broadcasting] over public opinion and the life of the nation'. Despite forceful counter-arguments that state control not only inhibited the growth of viewing choice but also dangerously concentrated 'the power of broadcasting over public opinion', and despite ingenious ideas (such as auctioning frequencies) for importing more market-oriented methods into the industry, little dent was made in official policy irrespective of the political party in power. Competition and choice in television have expanded only very slowly; 26 years elapsed between the ending of the BBC's monopoly in 1956 and the arrival of Channel 4 in November 1982.

Technology, however, develops at a faster pace than official thinking. The telecommunications revolution and advances in cable technology have suddenly awakened government to the potential—not least for jobs—of a whole new industry, centred on the television sets in people's homes, which promises not only to enlarge their access to entertainment, information and education but also radically to transform the nature of everyday activities like shopping and managing a bank account. Providing television by cable, rather than over the air, is not new; but, with the exception of a few local experiments, it has been tightly regulated to confine it to relaying existing broadcast television to remote outposts of the British Isles or to areas where, because of natural obstacles, over-the-air reception is poor. What is new is the development of sophisticated cables with a two-way ('interactive'), multi-channel capability which, allied to the computer and the television set, open up limitless horizons in the communications field. The growing potential of television as a component of this broader new industry in communications is making a self-contained policy for broadcasting increasingly unrealistic—and, indeed, damaging to the prospects for living standards. The new industry is unlikely to attract the private capital and entrepreneurship required for its success if it is shackled by the controls on cable imposed decades ago to protect the privileged position of broadcast TV.

In the spring of 1982, the Government—apparently rather abruptly alerted to the immense promise of cable systems—established a three-man Committee of Inquiry, under the chairmanship of Lord Hunt of Tanworth, with the following remit:

> 'To take as its frame of reference the Government's wish to secure the benefits for the United Kingdom which cable technology can offer and its willingness to consider an expansion of cable systems which would permit cable to carry a wider range of entertainment and other services (including when available services of direct broadcasting by satellite), but in a way consistent with the wider public interest, in particular the safeguarding of public service broadcasting; to consider the questions affecting broadcasting policy which would arise from such an expansion, including in particular the supervisory framework; and to make recommendations by 30 September 1982.'

The Committee met that tight deadline, and the Home Secretary in turn lost little time in informing the House of Commons, on 2 December, that the Government broadly endorsed Hunt's conclusions. Announcing a few firm decisions, he promised a White Paper early in 1983 setting out the Government's full response to Hunt and its detailed plans for cable TV.

In *Hobart Paper 96,* Dr Cento Veljanovski and Mr William Bishop subject to critical economic analysis the issues surrounding the expansion of cable TV systems. They also offer a detailed evaluation of the Hunt Committee's Report whose 'de-regulatory' stance they consider marks a watershed in broadcasting policy. They see the Government's decision to accept the general approach of Hunt as effectively jettisoning the principles upon which that policy has been based since the inception of the British broadcasting system 60 years ago. The era of 'rationed TV', they say, will soon be over.

In Section 1 of the *Paper* the authors set out the issues and the arguments for and against a liberal régime for cable expansion. They describe the technological challenge to traditional broadcasting policy and the reasons for the Government's desire to proceed urgently with cable. They make clear from the outset their lack of sympathy with the view that allowing cable TV to compete with the 'public broadcasting' system will impoverish the latter and lead to a fall in the standards of British television generally. Throughout the history of broadcasting in this country, they recall, new competition has always been portrayed as 'destructive' of whatever arrangements already existed, whether the threat has been to radio from television or to the BBC from commercial broadcasting. That such fears were not realised in the past suggests that the same argument currently being deployed against cable is special pleading designed not to further the public interest but to protect the vested interest.

In Section 2 the authors lucidly explain the technical features of supplying television by cable and describe how it differs from (and can complement) wireless broadcasting. They review the UK's experience to date with a variety of pilot schemes and the growth of pay-cable in the USA.

In Section 3 they argue that cable TV will be part of a wide and highly competitive product market including other forms of television, entertainment, education, and many types of communication. They also analyse the main reasons advanced for excluding the market from broadcast TV and find them deeply flawed, concluding that a market is feasible and could handle the technical problems held to justify the present public broadcasting system with its high hidden costs.

In Section 4 the authors argue that the *source* of financing television has a crucial effect on programme standards. They compare taxation (the BBC-type licence fee), advertising revenue and pay-TV, and conclude—as did Sir Sydney Caine in 1968—that the last is superior because it alone provides TV programmers with a direct measure of the preferences of viewers. They maintain that the case for pay-TV rests less on its ability to offer minority-taste programmes than on its responsiveness to consumer preferences. Commercial pay-TV will tend towards neither low-quality, nor high-quality, nor mass-appeal programmes, but towards the most profitable degree of variation in quality.

In Sections 5 and 6 the authors discuss the implications of the likelihood that, because laying cable requires a large fixed capital investment, economies of scale will justify only one cable system in a given area. Though analysing the principal alternative ways of controlling a local monopoly, they argue that cable TV is not a typical natural monopoly and requires neither public utility status nor extensive public regulation. They advance reasons for believing that, even if cable operators do enjoy a local monopoly, they are unlikely to be able to exercise significant market power.

Section 7 is devoted to a critique of the Hunt Report. The authors applaud it for recommending very few restrictions on cable to protect the BBC and ITV. Despite approving Hunt's general 'hands-off' approach, however, their lengthy examination of the Report is critical of a number of what they consider to be needlessly anti-competitive and potentially costly proposals. They also censure the Report for shallow analysis which, they claim, proceeds all too frequently by assertion rather than reasoned argument.

Finally, in Section 8, Dr Veljanovski and Mr Bishop outline the principles—and a supervisory framework to promote them—which *they* propose as the basis of government policy on cable. They maintain that pay-cable TV is akin to publishing because, unlike traditional broadcasting, it is in essence a private relationship between a subscriber and a cable operator. It is consequently free of the public interest concerns which have given rise to the principles of public accountability, objectivity and balance currently governing broadcasting. Moreover, the services supplied by pay-cable are in the nature of leisure, or luxury, goods. For both these reasons, neither public provision nor extensive public regulation of cable is desirable. Rather, because it gives full scope to individual initiative and experimentation and can adjust quickly and flexibly to changing demands, the competitive *market* should be used to the maximum feasible extent to finance and supply cable services. In particular, the authors argue against exclusive franchises which they consider will erect artificial barriers to entry incompatible with the promotion of competition. Further, they end with a call for serious consideration to be given to the case for de-regulating *broadcast* TV.

The Institute wishes to thank Sir Sydney Caine, Mr John Burton of the University of Birmingham, and Dr Brian Hindley of the London School of Economics for reading an early draft of this *Paper* and offering comments and criticisms which the authors have taken into account in their final revisions.

Although the constitution of the Institute requires it to dissociate its Trustees, Directors and Advisers from the authors' analysis and conclusions, it offers their *Hobart Paper* as an authoritative and timely contribution to the 'winter of debate' on the expansion of cable TV in this country.

January 1983 MARTIN WASSELL

THE AUTHORS

CENTO G. VELJANOVSKI was born in 1953 in Melbourne, Australia, and studied Law/Economics at Monash University where he graduated B.Ec. (Hon.) First Class in 1974 (M.Ec. 1976). He worked for the Australian Treasury, 1974-75, and was a Teaching Fellow in the Department of Economics at Monash University, 1975-76. He won a Commonwealth Scholarship to study in the UK and was awarded a D.Phil. by the University of York in 1981. He was a Visiting Lecturer at the University College at Buckingham, 1978-80; Visiting Professor in Law and Economics, University of Toronto, 1980-81. He has been a Research Officer in Law and Economics at the Centre for Socio-Legal Studies, University of Oxford, since 1978, and was elected to a Junior Research Fellowship at Wolfson College, Oxford, in 1979.

Dr Veljanovski specialises in the economics of government regulation, law and competition policy. He was joint editor (with P. Burrows) and a contributor to *The Economic Approach to Law* (Butterworths, 1981), and has contributed articles to law journals, including the *British Journal of Law and Society, Modern Law Review,* and *Law and Policy Quarterly,* and also to the *Scottish Journal of Political Economy* and the *Economic Journal.*

WILLIAM D. BISHOP was born in 1950 at St. Johns, Newfoundland, Canada, and educated at Memorial University of Newfoundland where he gained a BA, at the University of Western Ontario (MA (Econ.)), and at the University of Oxford (BA, BCL). He was a Lecturer in Law at Lincoln College, Oxford, 1975-76, and has been a Lecturer in Law at the London School of Economics since 1976. He was Visiting Professor, Law and Economics Programme, University of Toronto, 1979-80.

His research interests are in the application of economics to law, including the common law and competition law. He is the author of numerous papers in leading journals including the *Journal of Legal Studies, Modern Law Review, Law Quarterly Reveiw* and *Oxford Journal of Legal Studies.*

ACKNOWLEDGEMENTS

We are grateful to the Institute of Economic Affairs for providing us with this opportunity to write a *Hobart Paper* on the economic issues raised by cable regulation.

The editorial efforts of Martin Wassell and Michael Solly have not only earned our admiration but forced us to sharpen our analysis. For this we are grateful.

Finally, thanks are due to Stanley Besen, Hudson Janisch, Eli Noam, and Jacob Trobe for providing us with information on cable TV in North America.

C.G.V.

W.D.B.

ONE: The Winter of Debate

The eagerly-awaited Hunt Report on cable TV and broadcasting policy,[1] handed to the Government in September 1982, was produced in record time. It is the result of six months of deliberations by a three-man committee appointed to recommend to the Government how public service television broadcasting should be protected and cable TV regulated 'in the general public interest'. In proposing a relatively liberal 'supervisory framework' for the cable industry, the Report contains very little good news for broadcasters.

The Government must now formulate its policy towards cable in the light of the Report's recommendations—which mark a watershed in broadcasting policy—and the 'winter of debate'. The recent decision by the Government accepting its general approach has effectively jettisoned the principles upon which the British broadcasting system has been based since its inception over 60 years ago.[2] The era of 'rationed TV' subject to public operation and extensive regulation will be over. Cable will have a 'de-regulated status . . . unprecedented in British broadcasting practice'.[3]

1. The Issues

British television is confined to broadcast TV organised as a highly-regulated duopoly. The 'publicly'-operated British Broadcasting Corporation (BBC) offers two channels—BBC1 and BBC2. And, since 2 November 1982, two commercial channels, financed by advertising, have been provided by the independent television programme companies 'regulated' by the Independent Broadcasting Authority (IBA). To many people the high quality of television in Britain and its inter-

[1] *Report of the Inquiry into Cable Expansion and Broadcasting Policy,* Cmnd. 8679, HMSO, 1982. Subsequent reference to the Report will be made only by paragraph numbers.

[2] House of Commons *Hansard,* Vol. 33, No. 22, 2 December 1982, col. 417.

[3] Sean Day-Lewis, 'Go-Ahead Urged for Cable TV', *Daily Telegraph,* 13 October 1982.

national reputation as the best in the world is due to its being publicly regulated. In a nutshell, the case against cable TV can be stated thus: if it is allowed to compete with the public broadcasting system, the latter will be impoverished and the standards of British television generally will fall.

This argument is not new to the student of British broadcasting policy. Whether the threat has been to radio from television or to the BBC from commercial broadcasting, the new competition has always been portrayed as 'destructive' of the good world of broadcasting as then known. That the argument today is the same as in the past, the predictions of which did not prove accurate, strongly suggests that it is special pleading designed not to further the public interest but to protect the vested interest.

The old familiar arguments

The battle between the protagonists has been drawn up on old and familiar lines. The case against cable is built around the theme that unfettered competition is destructive and that commercialism is the enemy of good television. It is argued that cable will reduce the audiences and advertising revenues of broadcast TV and siphon-off the best programmes; that it is socially divisive, partitioning the country between those with it and those without it. It is claimed that cable will be available only in densely-populated, high-income areas and will thus re-inforce existing economic and social inequalities. As one submission to the Hunt Inquiry colourfully put it:

> 'If cable becomes symbolic of what Mayfair can have but Brixton cannot, what Metropolitan Man may enjoy but Rural Man is denied, then one more social tension will be generated in an uneasy age.'[1]

The opponents of cable TV maintain that, if the BBC and ITV are made to compete with it, their revenues and audiences will shrink and their ability to buy and produce good-quality programmes will be severely impaired. If programming is left to the logic of the market-place, they believe, there will be no diversity or innovation, no financially-risky programmes —only what Melvyn Bragg has described as the prospect of

[1] Cited in Alasdair Milne *et al.*, *The Cable Debate—A BBC Briefing*, British Broadcasting Corporation, London, 1982, App. III.

'a tyranny of mediocre monotoned similarity'.[1] The end result will be a 'programming wasteland'. Either that or cable operators will turn to the allegedly lucrative showing of pornography to recoup their massive investments. Furthermore, cable operators will enjoy 'local monopolies' enabling them to exploit consumers and to determine the programme content of an influential and persuasive means of communication. As the Association of Cinematograph, Television and Allied Technicians' (ACTT) submission to the Hunt Inquiry succinctly said:

> 'The power and immediacy of television as a means of communication, and the fact that the new cable networks will constitute local monopolies of their kind, remain as the basic justifications of the need for regulation.'[2]

In short, it is argued that, to ensure the vitality and principles of public service broadcasting and to maintain the current high standards, cable TV must be carefully regulated.

2. The Cable Goldrush?

Perhaps for the first time in British broadcasting history such arguments appear to have lost the day. The Government, by endorsing the recommendations of the Hunt Report, sees cable TV not as a threat to broadcast TV requiring detailed regulation, but as a 'supplementary' system widening the choice available to the British public.

The pro-cable lobby has been persuasive and has created a general euphoria about the benefits the expansion of cable will bestow. It has portrayed cable TV not only as ushering in a new—the third—age of broadcasting, but also as an economic and social wonderland which will bring untold blessings to the consumer and prosperity to the nation. The effects of cable have been alternatively compared to those of the Industrial Revolution, the Renaissance, the laying of the railway tracks in the 19th century, and the invention of the printing press.

Among the advantages to consumers claimed for cable are

[1] Melvyn Bragg, 'Goodbye Auntie, Enter the Multichannel TV Swapshop', *Sunday Times*, 5 September 1982.

[2] *Submission of the Association of Cinematograph, Television and Allied Technicians,* May 1982, p. 2.

more diversity and choice of programmes, and a closer matching of programmes with consumer preferences since minority tastes can be more easily catered for. Programmes will no longer be determined by the mass audience, advertisers or highbrow regulators, but rather by consumer demand in the market-place. Furthermore, cable will eventually form the basis of a nationwide communications system which will revolutionalise the way we live and work; it will bring the future closer and for that we should be grateful. Those who advise caution are belittled as Luddites, caricatured as a group of barons and monks with vested interests in the *status quo* who, were this the 14th century, would be solemnly debating whether we should have the Renaissance.[1] The notion that the Government should proceed *gradually* with cable expansion has been dismissed by one Minister as yet another example of the British 'genius for institutionalising torpor'.[2]

The above is a distillation of the views expressed so far in the debate about cable. Most exaggerate in one way or another; and some are sheer polemic. *All* require careful appraisal.

3. The Technological Challenge to Policy

For the past 60 years broadcasting policy in the UK has had one central premise: that broadcasting is a public service and should be provided by public authorities specifically set up to serve the public interest. The premise is based in part on the *technology* of broadcasting whose principal characteristic is that the electro-magnetic spectrum is limited. Over-the-air broadcasting uses scarce space on the spectrum which constrains the number of competing channels that can be offered to the public. Thus government regulation is necessary to allocate frequencies and prevent interference among different stations.

Although it does not follow that broadcasting must therefore be operated by government, that inference has been drawn and used to support a state monopoly—and, since 1956, a duopoly—in broadcast TV. The Sykes Committee stated in 1923 that 'The wavebands available in any country must . . . be regarded as a valuable form of public property'.[3] Government

[1] Peter Jay, cited in Melvyn Bragg, *op. cit.*

[2] Mr Kenneth Baker, Minister for Industry and Information Technology, cited in Bragg, *op. cit.*

[3] *The Broadcasting Committee: Report,* Cmd. 1951, HMSO, 1923.

control was further justified in the eyes of the Annan Committee by the 'potential power [of broadcasting] over public opinion and the life of the nation'.[1] In short, the scarcity of frequencies, the risk of signal interference and the considerable potential of broadcast TV to 'influence and offend' required it to be publicly controlled.

Cable, and the new technology of telecommunications, challenges this traditional philosophy and the argument for government regulation of TV. Cable is 'narrowcasting'; it has unlimited channel capacity. It has been likened to publishing and therefore, it is claimed, requires no more controls than those to which publishing is subject.

Even objectivity, a fundamental principle of the public service broadcasting system, has no necessary place in cable TV. A fully-fledged cable system offers discretionary viewing paid for on a per-channel basis with a large choice of different channels. Cable TV, like publishing, should therefore be allowed to reflect the different views, ideologies, prejudices and tastes of the British public.

A new policy for broadcasting has been necessitated by the rapid developments in technology which are breaking down the divisions between different segments of the communications industry. The decision to allow cable expansion is but one manifestation of how technology has undermined the principles and philosophy underlying policy in communications. Indeed, it is becoming increasingly unrealistic to think in terms of a self-contained policy for broadcasting. Cable and broadcast TV are becoming part of a wider industry in communications and will in the future compete with other media in supplying information ('Prestel' and 'Teletext', for example), entertainment, and a wide range of communication services.

4. Why the Urgency? Jobs and Exports

After years of neglect, telecommunications has been given a high priority in the industrial policy of the Thatcher Government. In the last two years a number of significant initiatives have been taken; a virtual public monopoly in telecommunications is being opened up to private enterprise. In the catch-

[1] *Report of the Committee on the Future of Broadcasting* (Annan Report), Cmnd. 6753, HMSO, 1977, para. 3.2.

words of the Government, it is being 'privatised' and 'de-regulated'. British Telecom's telephone service and equipment monopolies have been dismantled and are being privatised.[1] A new broadcasting TV station (Channel 4) is now on the air and two breakfast TV channels have also begun transmitting. In addition, the Home Office has agreed to the introduction of direct broadcast by satellite (DBS) TV which will provide a further two channels,[2] and has licensed 13 experimental pay-cable TV stations. The Prime Minister has already given the green light to a 30-channel interactive cable system.

The reason for the Government's sense of urgency can be summed up in the words 'jobs' and 'exports'.[3] Cable and satellite expansion are part of an attempt to revitalise the economy and place UK electronic industries at the forefront of technological development in telecommunications in Europe.[4] The authors of the much-publicised Cabinet Office report entitled *Cable Systems* recommended a 30-channel interactive cable system which they estimated would require a capital investment of £2,500 million and which would stimulate 'subsequent economic activity . . . of the order of £1,000 million annually', bringing benefits to British industry and workers.[5] Britain already has a comparative advantage in satellite technology. British Aerospace is the leading manufacturer of satellites in Europe and has led the world in the development of large direct broadcast satellites. These facts were influential in the Government's decision to give the

[1] For those not familiar with the term, 'privatisation' means the selling to the private sector of 51 per cent or more of the shares in a public corporation. Privatisation may, however, take many forms. (John Redwood and John Hatch, *Controlling Public Industries*, Basil Blackwell, Oxford, 1982, Chap. 7.)

[2] Britain has been given five DBS channels. The Home Secretary has decided to allocate two channels immediately to the BBC and the other three 'when demand is justified'. (Home Office, *Direct Broadcasting by Satellite*, HMSO, London, 1981.)

[3] Mr Kenneth Baker, the Minister for Industry and Information Technology, confirmed this interpretation in the recent House of Commons debate: 'The reason we want to move quickly is [that] with cabling more jobs will be created'. (*Hansard, op. cit.*, col. 494.)

[4] Similar, though less rapid and extensive, developments are taking place across the Channel. In West Germany four government-operated pilot cable TV schemes have been authorised. In France a socialist government is lifting the TV programming monopoly and is about to launch the country's first large-scale cable TV system with a target of 1·5 million households wired by 1986. On France, David White, 'French tie plans for cable TV to fibre optics technology', *Financial Times*, 14 October 1982.

[5] Cabinet Office, *Cable Systems*, HMSO, 1982, para. 8.

go-ahead for two satellite TV channels to be operated by the BBC.

Cable expansion is therefore perceived as a source of new wealth for the nation (and of electoral success for the Government!). Mrs Thatcher's decision to wire Britain 'as a matter of urgency' has been described as 'the most important industrial decision of her administration'.[1]

Genesis of a policy

Given this background, it is easy to understand why the Hunt Report is so favourably disposed to cable TV. Indeed, the Committee of Inquiry was precluded from considering the broader issues in being asked to 'take as its frame of reference the Government's wish to secure the benefits for the United Kingdom which cable technology can offer'. The sense of urgency with which policy is being formulated and the issues are being debated has, however, given rise to concern in some quarters. Peter Fiddick, writing in the *Guardian*, has described the Hunt Report as 'part of a political process so swift and so foggy as to be unprecedented'.[2] Some fear that the Government's rush to formulate a cable policy may lead to the adoption of an ill-conceived regulatory framework. This danger is seen to be confirmed by the Hunt Report which contains no thorough or systematic analysis of the issues raised by cable expansion.

The significance of the Hunt Report is its 'de-regulatory' stance, rejecting the view that television should be the exclusive preserve of the BBC and ITV. For the last 60 years successive governments, the Conservatives included, have protected public broadcasting from competition; cable TV has been heavily regulated. Except for a few minor experiments, cable TV companies have been permitted to relay only BBC and ITV programmes—in full and simultaneously. As a result, the market for the services of cable has been confined to those areas where the reception of broadcast BBC and ITV is poor.

Earlier inquiries into broadcasting saw no reason to disturb this arrangement. The last, the Annan Inquiry, could find no justification for pay-cable TV to compete with existing

[1] 'The Wiring of Britain', *Economist*, 6 March 1982, p. 11.

[2] Peter Fiddick, 'Rescuing cable television from its shameful beginnings', *Guardian*, 18 October 1982.

broadcast TV.[1] Its report claimed that pay-cable would not generate new programme material and 'was therefore a ravenous parasite. It lived off those who produced television and films.'[2] Annan recommended, instead, that cable TV should develop as a local community service. Community cable would not duplicate broadcast TV but would confer the double benefits of extending the number of programme-makers and giving the man-in-the-street an opportunity to make his own programmes.

Labour's tentative break with traditional policy

The first change in policy came with the 1978 White Paper on Broadcasting (the colour indicating an official statement of government policy). Unlike Annan, the then Labour Government was not prepared 'to dismiss the possible advantage of pay-TV', nor to conclude that the alleged disadvantages of cable were insurmountable. The answer, it felt, might be found by 'experimentation'. The White Paper stated that, 'in principle, there seems to be no reason why both pay television and community cable systems should not develop side by side'.[3] It outlined the Labour Government's policy:

(a) New legislation to permit pilot pay-cable TV schemes subject to 'careful regulation to guard against the possibly damaging effects which pay-TV might have on television as a whole, and on the cinema industry';[4]

(b) The transfer of the licensing and supervision of cable TV from the Home Office to the IBA.[5]

The Home Office licensed 13 pilot cable schemes in 1981, but recommendation (b) has not been implemented.

In March 1982 the Information Technology Advisory Panel (ITAP) of the Cabinet Office published a report entitled *Cable Systems.*[6] ITAP is a group of advisers from the electronics industry appointed by the Prime Minister 'to ensure that govern-

[1] Annan Report, *op. cit.*, paras. 14.34-14.56.

[2] *Ibid.*, para. 14.50.

[3] *Broadcasting*, Cmnd. 7294, HMSO, 1978, para. 175.

[4] *Ibid.*, para. 178.

[5] *Ibid.*, para. 181.

[6] *Op. cit.*

ment policies and actions are securely based on a close appreciation of market needs and opportunities'. Its report recommended that:

(i) the present restrictions on cable programming should be removed;

(ii) a new statutory body should be formed to regulate the industry;

(iii) the industry should develop an effective system of self-regulation;

(iv) the Department of Industry should set up a working group to develop technical standards;

(v) the Government should announce the broad outlines of its future policy by mid-1982.

ITAP report's theme of de-regulation passed on to Hunt

The theme of the ITAP report, not surprisingly, was that the Government should not hesitate to de-regulate the cable industry. It stated its opinion:

> 'that only through a set of speedy, positive and radical regulatory changes can the United Kingdom obtain the benefits offered by developments in the cable technology . . . For British industry a late decision is the same as a negative decision.'[1]

This 'decide now' imperative immediately became a feature of the Government's approach.

Within days of the publication of the ITAP report, the Home Secretary announced the appointment of Lord Hunt to head a three-man Committee of Inquiry into the broadcasting aspects of cable.[2] The Inquiry was asked by the Home Secretary:

> 'To take as its frame of reference the Government's wish to secure the benefits for the United Kingdom which cable technology can offer and its willingness to consider an expansion of cable systems which would permit cable to carry a wider range of entertainment and other services (including when available

[1] *Ibid.*, para. 8.13.

[2] In addition to Lord Hunt, an ex-civil servant and former Secretary to the Cabinet, the other members of the Committee were Sir Maurice Hodgson, formerly chairman of ICI and currently chairman of British Home Stores, and Professor James Ring, Professor of Physics at Imperial College, London, and an ex-member of the IBA.

services of direct broadcasting by satellite), but in a way consistent with the wider public interest, in particular the safeguarding of public service broadcasting; to consider the questions affecting broadcasting policy which would arise from such an expansion, including in particular the supervisory framework; and to make recommendations by 30 September 1982.'[1]

On 7 April 1982 the Committee issued a consultation document, asking a number of specific questions, which it distributed to interested parties.[2] The questions related mainly to restrictions on the ownership and financing of cable, programme standards, protection of the cinema, and the form and nature that regulation of the industry should take. The Committee received 189 written submissions.

While the Committee was working, the *Sunday Times* leaked Mrs Thatcher's decision to give the 'go-ahead for a 30-channel cable system'.[3] The Hunt Report was subsequently handed to the Government one day before the tight deadline of 30 September, incorporating its authors' view that they had been given 'sufficient [time] to allow us to properly consider the issues'. That view, it must be said, is not confirmed by a reading of the Report.

A less publicised inquiry has also been undertaken by the Department of Industry on technical standards and the role of British Telecom (BT) and Mercury in the cable industry. These non-TV aspects of cable policy, about which little is known, are of equal importance to the industry.

The Home Secretary has since announced that the Government is 'broadly in accord with the general approach and particular recommendations of the Hunt Report'.[4] A White Paper will be published early in 1983—possibly in February. Thus the decision to cable Britain and the formulation of public policy on it have taken little over a year. To put this in perspective, it should be recalled that it took 18 years for a government finally to agree to a fourth broadcast TV channel. This assessment may, however, prove too optimistic since implementation will depend on the result of the next General Election which may take place in 1983.

[1] Reprinted in Appendix A of the Hunt Report.

[2] *Ibid.*

[3] Richard Brooks, '30-channel TV gets Thatcher go-ahead', *Sunday Times,* 29 August 1982.

[4] House of Commons *Hansard,* 2 December 1982, *op. cit.,* col. 417.

TWO: Cable Technology and Potential

The general excitement surrounding cable television must be understood in the context of the rapid technological advances now taking place in telecommunications. In this sphere, more than in most others, public policy must be grounded in a full understanding of the technical features of the 'product'. This Section gives a brief account of the technology of cable, its potential uses, and the rapid growth in pay-cable which has taken place in the USA since the industry there was de-regulated.

1. Cable Technology

A cable system is simply a way of carrying video and audio signals into homes and other locations by a wire—much like the telephone. That is the most visible difference between cable and conventional broadcasting which transmits through the atmosphere and is received by a rooftop aerial.

All cable systems have three basic features: the 'headend', the cable network, and the subscriber terminal.

(i) *The headend*

The headend is the nerve centre of a cable system. It consists of a number of aerials and other devices which receive signals from broadcast TV, satellites and local production studios. In the case of broadcast TV, the headend receives the signal from a tall mast aerial situated on a high point of land. Similarly, satellite signals are received by parabolic dish antennae pointed at the sky in the direction of the satellite.

The headend re-transmits these signals along a wire to the subscriber. Each signal represents a channel of programmes. By increasing the number of receivers at the headend, a cable operator can offer the subscriber more channels. The channel capacity depends on the type of cable used but is potentially unlimited.

Most cable systems permit the subscriber to receive only

incoming signals, that is, they are one-way communication systems like broadcast television. Some of the newer systems, however, are 'interactive', permitting two-way communication between the subscriber and the headend and, through the latter, between subscribers. In an interactive cable system the subscriber can transmit his own signals using a keyboard attached to his TV set. To date, only one per cent of cable systems in the United States are interactive.

(ii) *The cable*

The second feature of a cable system is the cable itself. It is a wire which carries the signals from the headend to the subscriber's terminal. It is either buried under the ground, ducted or strung overhead.

The network can be constructed of coaxial or optic fibre cable, or a combination of both. A coaxial cable consists of an inner copper wire of small diameter which passes through a copper mesh of larger diameter from which it is separated by plastic foam. The mesh is in turn entirely sheathed in a plastic outer cover to protect the copper wire from weather and electrical interference. The newer coaxial cables have a capacity of 50 channels, though the number can be increased almost without limit by laying more separate cables.

Optic fibre cable is much newer and technically superior to copper wire for the transmission of signals. Optic fibres are exceedingly pure, hair-like threads of glass. They are as strong as steel but flexible (they can be tied into a knot without snapping). Unlike coaxial cable, which transmits signals by electrical impulses, optic fibres use bursts of light. The major technical attractions of optic fibres are their ability to carry signals over longer distances than coaxial cable and their immunity from electrical interference.

Another important attribute of optic fibres is their almost unlimited channel capacity. A relatively thick, twisted copper wire (such as is currently used for local telephone calls) can, for example, carry at most 24 simultaneous telephone conversations. Two thin filaments of glass can carry 12,000. Scientists foresee a future where a single fibre may be capable of carrying up to 200,000 simultaneous conversations linking 400,000 subscribers.

Because of the high cost of optic fibre cable today, it is uneconomic to use it extensively. Currently, its principal application is in the main trunk line of the cable system. Technological development and growing markets, however, can be expected to bring down its cost in the near future.

The wire network of a cable system can be laid in one of two patterns. The most common resembles a tree. A trunk-line cable is laid from the headend and is branched out at intervals to serve groups of subscribers, and finally individual ones. Alternatively, the cable can be laid in a 'star system' with a number of trunk cables feeding 'exchanges', which are simply boxes on a wall, whilst a separate cable links the customer to the exchange.

(iii) *The subscriber terminal*

The final main feature of a cable system is the subscriber terminal which connects the TV set to the wire. The subscriber selects the channel he wishes to watch by turning a tuning 'dial' on his set in exactly the same way as at present.

2. PAY-CABLE

This *Hobart Paper* is principally concerned with one aspect only of the cable industry, subscription or pay-cable TV—that is, where the viewer/subscriber pays a monthly charge for his consumption of television.[1] Pay-TV is to be distinguished from broadcast TV which is 'free' at the point of consumption and financed by either a 'lump sum' tax (the licence fee) or advertising revenue.

Pay- or subscription-TV is not confined to cable. It is possible, although expensive, to provide broadcast and satellite TV on a pay-per-channel basis. One of the satellite TV channels to be operated by the BBC, for example, will be on subscription.[2] Nonetheless, it is easier and more cost-effective at present to provide pay-TV by cable. In the United States, cable has over 87 per cent of what is the biggest pay-TV market in the world (Table I).[3]

[1] Sydney Caine, *Paying for TV?*, Hobart Paper 43, IEA, 1968.

[2] *Direct Broadcasting by Satellite—Report of a Home Office Study,* HMSO, London, 1981.

[3] As Table I shows, however, subscription broadcast TV (or STV, as it is called in the industry) is beginning to grow rapidly. Also, Burt Schorr, 'Over-the-Air Pay TV Builds Subscribers, May Give Cable and Networks Headaches', *Wall Street Journal,* 25 September 1979.

TABLE I

GROWTH OF PAY-TV IN USA, 1979-80

Type	*No. of Subscribers*	*Increase 1979-80*	*Share of total pay-TV market*
	'000	%	%
Pay-cable	8,700	51·8	87·7
Subscription Broadcast Television	825	106·5	8·3
Multipoint Distribution System	400	44·1	4·0

Source: Eli M. Noam, 'Cable Television, A Statistical Overview', Research Working Paper No. 448A, Columbia University, 1981.

Cable TV is marketed in tiers. A subscriber connected to cable receives a number of channels and programmes for his basic monthly subscription fee. This is often referred to as the 'basic tier', or simply the 'basic'. In Britain it might consist of BBC and regional ITV, some foreign broadcast television and a number of channels with general appeal offering, for example, news and children's programmes. Pay-cable refers to channels or programmes which are in addition to the basic service and for which an extra monthly fee is charged. Such 'premium' channels will offer, for example, continuous films, sport, general or financial news, arts, and public access and community cable programmes.[1]

Methods of charging for pay-cable

There are two methods of charging for pay-cable—by channel or by programme. Pay-per-channel pricing is almost universal because it is relatively cheap.

(i) *Pay-per-channel*

Pay-per-channel TV systems use a variety of devices to distinguish paying from non-paying subscribers. The cheapest and most common is the 'negative trap' which is attached to the exterior of the house of the non-paying subscriber and

[1] *Public access* channels are reserved on a first-come, first-served basis for individuals and groups who are not professional programme producers or distributors. *Community cable* provides a means of communication between members of a local community to discuss local affairs and politics and for local events to be televised.

blocks reception of the pay-channel. If a subscriber wants to receive the signal, he asks the cable company to send a technician to remove the trap. Frequently, the labour required to install and remove the trap costs more than the device itself. Moreover, traps give rise to a security problem; they can be tampered with relatively easily. It is estimated that around 25 per cent of cable subscribers in the United States receive pay-channels free of charge.[1]

(ii) *Pay-per-programme*

Pricing cable TV by programme—called 'pay-per-view' by Hunt—is uncommon for a simple reason. Whereas in a pay-per-channel system it is necessary only to distinguish paying from non-paying viewers, pay-per-view requires a record to be kept of each programme watched by each subscriber. This additional metering function explains why pay-per-view technology is relatively more complex and therefore more expensive. To be feasible and effective, paying by programme requires an interactive cable system enabling the operator to monitor the subscriber's viewing by computer. Network One in Toronto, for example, uses two-way cable to meter the programme consumption of its subscribers. The subscriber is given a 10-minute preview of the programme, after which he must push a button if he wishes it to continue. He receives an itemised bill at the end of the month.

3. The Attraction of Cable

In one sense, the notion that cable TV marks a technological advance appears strange. The stringing of wires across Britain to enable people to receive entertainment and information smacks of horse-and-buggy technology in a satellite age. To appreciate the potential of cable, however, it is necessary to understand how it differs from (and can complement) wireless broadcasting.

(i) *Almost unlimited channels*

Cable TV has the capacity for a large number of channels, whereas both conventional broadcast TV and DBS are constrained to a handful of channels because of the scarcity of

[1] 'Two-way Stretch', *Economist*, 5 June 1982, pp. 62-3.

frequencies.[1] Some of the newer cable systems in the USA, for example, can handle 100 channels. Cable TV has been aptly described as the 'television of abundance'.[2]

(ii) *Better picture*

Cable TV also produces a picture of superior quality. The cable protects the signal from interference by the weather and physical obstacles such as mountains and tall buildings. Thus its subscribers are spared the irritation of 'shadows' and other symptoms of poor reception; the picture they get is usually clearer and more defined than that received through the air.

(iii) *Narrowcasting*

A third attraction of cable is that it is local. It is *narrowcasting,* delivering its programmes only to those who want to subscribe in each locality served by the cable system.[3] Together with its capacity for a large number of channels, this gives the subscriber a wider choice and permits minority-interest programmes to be shown. In the United States, cable operators offer a broad range of services including continuous sport, uninterrupted and 'adult' films, ethnic, educational and local-interest programmes, 24-hour news, classified advertisements, and general information services. Pay-cable enables TV both to cater more finely for the diverse preferences of television viewers and to gauge the intensity of their preferences for different services.

[1] Britain has been allocated five DBS channels (by the World Broadcasting—Satellite Administrative Radio Conference held in Geneva in 1977), two of which will come into operation in 1985. DBS requires the satellite to be 'parked' 22,200 miles above the equator so that the time it takes to orbit the earth is exactly equal to the time it takes the earth to rotate around its axis. The satellite thus appears stationary when viewed from earth. In order to avoid signal interference, satellites must be parked some distance from each other. A recent study estimates that DBS will be able to offer between 4 and 10 channels per satellite: Staff Report, *Policies for Regulation of Direct Broadcast Satellites,* Federal Communications Commission, Washington, DC, September 1980. Also, *Direct Broadcasting by Satellite,* HMSO, London, 1981, Chs. 2 and 3; and Barry Fox, 'Satellite TV starts the Ultimate Craze', *New Scientist,* 7 September 1982.

[2] Sloan Commission on Cable Television, *On the Cable: the Television of Abundance,* McGraw Hill, New York, 1979.

[3] *Narrowcasting* is defined by Hunt as the 'ability to provide television on a truly local scale and in such abundance that all kinds of specialist interests could find themselves catered for in a way that is not possible on broadcast television' (para. 65).

(iv) *Complements broadcast TV*

Fourthly, cable systems complement broadcast TV. In most countries cable began as a relay service for over-the-air broadcasting in areas with poor reception; and it can bring distant and foreign television to the subscriber. It can also be used to build a national distribution network for telecommunication satellites in space and as a general communications system.

That cable TV can extend other forms of television does not, of course, mean that it does not compete with them. The nature of the competition, however, depends crucially on both the economic realities of supply and demand and the public regulation of cable.

(v) *Non-television uses*

It is generally accepted that cable expansion in the UK will be primarily in response to the demand for more entertainment. (In the USA, 80-85 per cent of the material shown on cable TV consists of feature films.) Yet cable has a myriad of non-entertainment uses which may in the long run prove more important. The two-way version will permit subscribers to use cable as an efficient and integrated communications system. The futurologist's vision of a 'wired city' where cable facilitates many of life's transactions is frequently invoked, and it is no doubt behind the Government's enthusiasm for cable expansion.[1]

Although this potential has not yet been developed because of its high cost, the technology now exists for people to shop, bank, sell, buy and even work using cable. Cable can also be adapted to read meters, to operate security and fire-alarm systems, and to provide general information, educational services and classified advertisements. Although these uses may develop as cable matures, it is clear that entertainment is—and will remain for some time to come—its main and most profitable service.

[1] This potential is recognised by Hunt: 'It seems to be generally agreed that investment in cable television for entertainment purposes will be the *necessary base* to which the interactive services of economic benefit to business and the individual will be added' (para. 5). (Italics added.) It is certainly the DoI's motivation in supporting rapid cable expansion. The Minister, Mr Kenneth Baker, stated in the Commons debate that 'the range of new non-broadcasting services is the *raison d'être* for the expansion'. (*Hansard,* Vol. 33, No. 22, *op. cit.,* col. 495.)

4. Cable as a Distribution System

The extent to which these potential advantages are realised will depend crucially on the financial realities of cable TV. A rather misleading view in the current public discussion is that cable will instantly grow and provide an abundance of channels. A more reasonable assessment would be that cable programming will expand gradually. It is a capital-intensive, high-risk industry and the demand of the British public is unknown. The range and number of programmes and other services provided by cable will depend not only on consumer demand but on its competitiveness *vis à vis* other methods of distributing entertainment and information. Cable, like broadcast TV, the cinema and other media, is simply a *distribution system* for communications. Cable will therefore enter an increasingly competitive market-place where it will have to provide a range of services at prices that compare favourably with those of other media. Otherwise it will not survive.

5. Cable in the UK

In 1924 in the village of Hythe near Southampton, Mr A. W. Morton used a wire to attach a radio receiving set to a loudspeaker in another room to enable his wife to listen there. By 1925 he had extended his wire radio system to 25 subscribers, charging each 1s. 6d. per week. Such was the beginning of the British cable relay industry, which expanded during the 1930s and 1940s.[1]

In the post-war years the idea was extended to television. But, although cable was technically capable of providing many more channels, the terms of the licences issued by government allowed cable companies to distribute only BBC and ITV programmes. With minor exceptions, the services offered by cable TV in Britain have been confined to the relay of broadcast TV in areas of poor reception. Government policy has been geared to achieving a nationwide coverage for broadcast TV. Since reception of the latter has improved, however, such that BBC and ITV broadcasts today reach 99 per cent of the population, the demand for cable relay services

[1] As recounted by Ronald H. Coase, *British Broadcasting: A Study in Monopoly*, Longmans Green for the LSE, London, 1950, pp. 69-70.

TABLE II

HOUSEHOLDS RECEIVING COMMERCIAL AND BRITISH TELECOM CABLE TV IN BRITAIN

(through private commercial companies and British Telecom only)

Year	*Households 000's*
1975	1,652
1976	1,624
1977	1,612
1978	1,577
1979	1,575
1980	1,483
1981	1,420

Source: Home Office.

has fallen. There has consequently been a decline in the number of subscribers (Table II).

In 1980 the Home Office licensed 2,291 cable operators to relay broadcasting services. About 2·6 million households, or 14 per cent of those with television sets, currently receive television by cable. Over half are serviced by private cable companies and British Telecom for which they are charged a small annual fee of the order of £15. The rest receive it through systems operated by local authorities, housing associations and community groups which often take the form of no more than a master aerial connected to all the TV sets in, say, a block of flats. Since these systems require a Home Office licence, official statistics considerably exaggerate the size of the relay cable business in Britain. Most of the systems are outdated, using twisted pair copper wires, and many have the capacity to carry only four TV channels.

6. Experiments in Cable Programming

To date there have been only three experiments in Britain allowing cable TV companies to provide their own programmes.

(i) *Pay-TV Ltd. in the 1960s*

Pay-cable TV had a brief and unsuccessful trial in the 1960s. Between 1966 and 1968, Pay-TV Ltd. (a private cable

TABLE III

COMMUNITY CABLE DEVELOPMENT IN FIVE ENGLISH CENTRES, 1972-82

Station	*Period*	*Houses wired*	*Subscribers*	*Capital cost*
		No.	*No.*	£
Greenwich Cablevision	July 1972 to present	20,000	12,000	20,000
Bristol Channel	May 1973 to March 1975	60,000	23,000	70,000
Sheffield Cablevision	October 1973 to January 1976	98,899	35,547	57,130
Swindon Viewpoint	n.a.	23,000	12,000	46,892
Cablevision Wellingborough	March 1974 to March 1975	n.a.	4,500	60,000

Source: Peter N. Lewis, *Community Television and Cable in Britain,* British Film Institute, London, 1978.

company) offered 'coin-in-the-slot' television in Southwark, Westminster and Sheffield to a total of 9,800 subscribers. The company lost money and attempted to persuade the then Labour Government to extend the experiment to at least 250,000 homes. The Government refused and the company closed down in October 1968.[1]

(ii) *Community cable*

In 1972 five community cable stations were licensed to distribute locally-produced programmes designed for a local community. The stations[2] were located in Greenwich, Bristol, Sheffield, Swindon and Wellingborough and had an estimated 87,000 subscribers (Table III). The original licences expired in 1976 but were extended to 1979.

[1] Annan Report, Cmnd. 6753, *op. cit.*, para. 14.35.

[2] For a discussion of these stations, Peter M. Lewis, *Community Television and Cable in Britain,* British Film Institute, London, 1978. Also, Annan Report, Cmnd. 6753, *op. cit.,* paras. 14.36-14.38.

These community cable experiments have been described as the 'video equivalent of the local newspaper'.[1] Under the conditions attached by the Home Office on issuing the licences, the stations were not, however, permitted to advertise or be sponsored, and they were required to offer subscribers programmes designed specifically for the local community. The Home Office removed the ban on advertising in 1975, but only two of the five stations took advantage of the change.

The stations provided both relay TV and radio *and* local-interest programmes, and offered, on average, up to 14 hours of television a week, half devoted to original programmes. They also were unsuccessful and all but one closed down for lack of funds.

(iii) *Pilot subscription-cable*

The third, and current, experiment is in pay-cable TV. Following the recommendations of the 1979 White Paper, the Home Office has licensed 13 pilot subscription-cable TV schemes. Applications for the licences were sought only from licensees of existing cable TV relay systems. In addition to relaying broadcast television, the successful applicants are now permitted to offer other programmes, such as feature films, at an extra charge to the subscriber. Under the terms of the licence, however, feature films may only be shown one year or more after they have been registered for public exhibition in a cinema, and no 'X'-rated film may be screened before 10 p.m. Advertising is not permitted; nor may a licensee seek the exclusive right to sporting and entertainment events of national importance. Operators' programming schedules must be submitted to the Home Office in advance.

The first of the 13 pilot subscription-cable TV schemes began operations in October 1981. They are listed, together with their programme suppliers and locations and the number of homes they pass,[2] in Table IV. At the end of 1981, they passed approximately 330,000 homes, 110,000 of which—one-third—had cable TV connected. Approximately 10 per cent of connected households subscribed to the pay-TV channel.

[1] *Cable Systems, op. cit.*, para. 1.4.

[2] Homes 'passed' refers to the number of homes which potentially could be linked up to a particular cable system. They are to be distinguished from homes which are actually connected to and receiving cable TV. The ratio of the latter to the former is often referred to as the 'penetration' or 'density' rate.

TABLE IV

PAY-CABLE TV SYSTEMS IN BRITAIN, 1982

Licensed operator	*Programme supplier*	*Location*	*Number of homes passed**
Rediffusion	Rediffusion	Reading	16,500
		Pontypridd	5,000
		Hull	8,000
		Tunbridge Wells	8,500
		Burnley	19,600
Radio Rentals	Thorn-EMI Video Productions	Swindon	22,000
		Medway Towns (Chatham, Gillingham and Rochester)	22,000
British Telecom	SelecTV	Milton Keynes	18,000
Philips Cablevision	SelecTV	Tredegar	6,000
		Northampton	6,000
Visionhire Cable	BBC Enterprises	London (various areas)	170,000
Cablevision	SelecTV	Wellingborough	8,000
Greenwich Cablevision	Greenwich Cablevision	Greenwich	20,000

Source: Home Office. *For definition of this term, p. 37, note 2.

To take one example, the Greenwich cable network has the capacity for 10 television and 40 radio channels. In television, it currently relays BBC1, BBC2, ITV London, ITV Anglia, one pay-TV channel (called 'Screentown'), Analogue Teletext, and now the new ITV Channel Four. Eight radio stations are also offered, including one pay-radio channel. Greenwich Cablevision imposes a basic charge for its relay services of £2.30 a month. For connection to the pay-channel 'Screentown', which provides films (including 12 new feature films), subscribers must pay an additional monthly sum of £9.20.[1]

The original purpose of the pilot schemes was to test public demand for pay-cable TV and the effects on existing broadcast television and the cinema industry. The licences were to run for two years, after which the Home Secretary proposed to review the situation to determine whether the schemes should continue and expand on a permanent basis—and, if so, what arrangements should be made for their supervision and operation. The Home Secretary also planned to formulate policy towards the film industry and any safeguards that might be needed to protect the cinema and broadcast television. These experiments in pay-cable TV have obviously been overtaken by recent events.

7. Growth of Pay-cable in the USA

The United States has the biggest cable TV industry in the world and is often looked to for indications of the potential effects of cable in the UK. The rapid growth of pay-cable in the USA after its de-regulation no doubt influenced the British Government's desire to free the UK cable industry.

The first cable system in the USA began operations in 1949. Like the industry it subsequently spawned, it simply relayed broadcast TV transmissions to households in mountainous and rural areas with poor reception. A master aerial was erected on a mountain top and signals were distributed to households in the valley by a coaxial cable. Hence the industry was usually referred to as 'community antenna television' (or CATV).

FCC intervention and subsequent de-regulation

Until the mid-1960s cable TV was relatively free of government regulation. In 1966, however, the Federal Communications Commission (an independent regulatory body responsible for communications) introduced a series of measures retarding the growth of cable TV in order to protect broadcast

[1] The Radio Rentals pay-channel ('Cinematel') is taken by 6 per cent of homes passed by the cable and 15 per cent of the homes cabled. A charge of £7.75 a month is made for the service. (P. Gosling's speech delivered to the Stock Exchange's *Cable Television Investment Seminar,* July 1982.)

TABLE V

GROWTH OF PAY-CABLE IN USA, 1973-80

	Pay-cable systems		*Pay-cable subscribers*	
	Estimated numbers	*As percentage of all cable systems*	*Estimated numbers*	*As percentage of all cable subscribers*
		%		%
1973	10	n.a.	18,400	n.a.
1974	45	1	67,000	1
1975	75	2	265,000	3
1976	253	7	794,000	7
1977	459	12	1,174,000	9
1978	790	21	2,352,000	18
1979	831	38	4,334,000	25
1980	2,000	48	7,500,000	50

Source: Barry Litman and Susanna Eun, 'The Emerging Oligopoly of Pay-TV in the USA', *Telecommunications Policy,* Vol. 5, June 1981, pp. 121-35, Table I.

TV from competition.[1] This policy came under increasing criticism, and after the mid-1970s many of the more restrictive regulations were removed. By 1978 the FCC was well on the way fully to de-regulating the industry.

De-regulation has led to a rapid growth in pay-cable in the USA. Since the first pay-cable service was offered by Home Box Office in 1972, it has become a big and varied industry offering a diverse range of programmes. Its growth can be clearly seen from Chart 1 and Table V.

In the USA there are 4,200 operating cable systems serving approximately 15 million subscribers, or nearly 44 million people, equivalent to 20 per cent of all households in the nation with television.[2] Most of these systems offer 12 channels. The cost of stringing overhead cable ranges from $8,000 a mile in rural areas to $12,000 a mile in urban areas. Laying cable underground is much more expensive and can amount to $80,000 a mile. Equipment costs are generally low; a small

[1] For a fuller account of the history of the FCC's efforts to regulate cable, Stanley N. Besen and Robert W. Crandall, 'Deregulation of Cable Television', *Law and Contemporary Problems,* Vol. 44, 1981, pp. 77-124.

[2] These statistics are for 1980 and come from *Broadcasting Yearbook, 1980.*

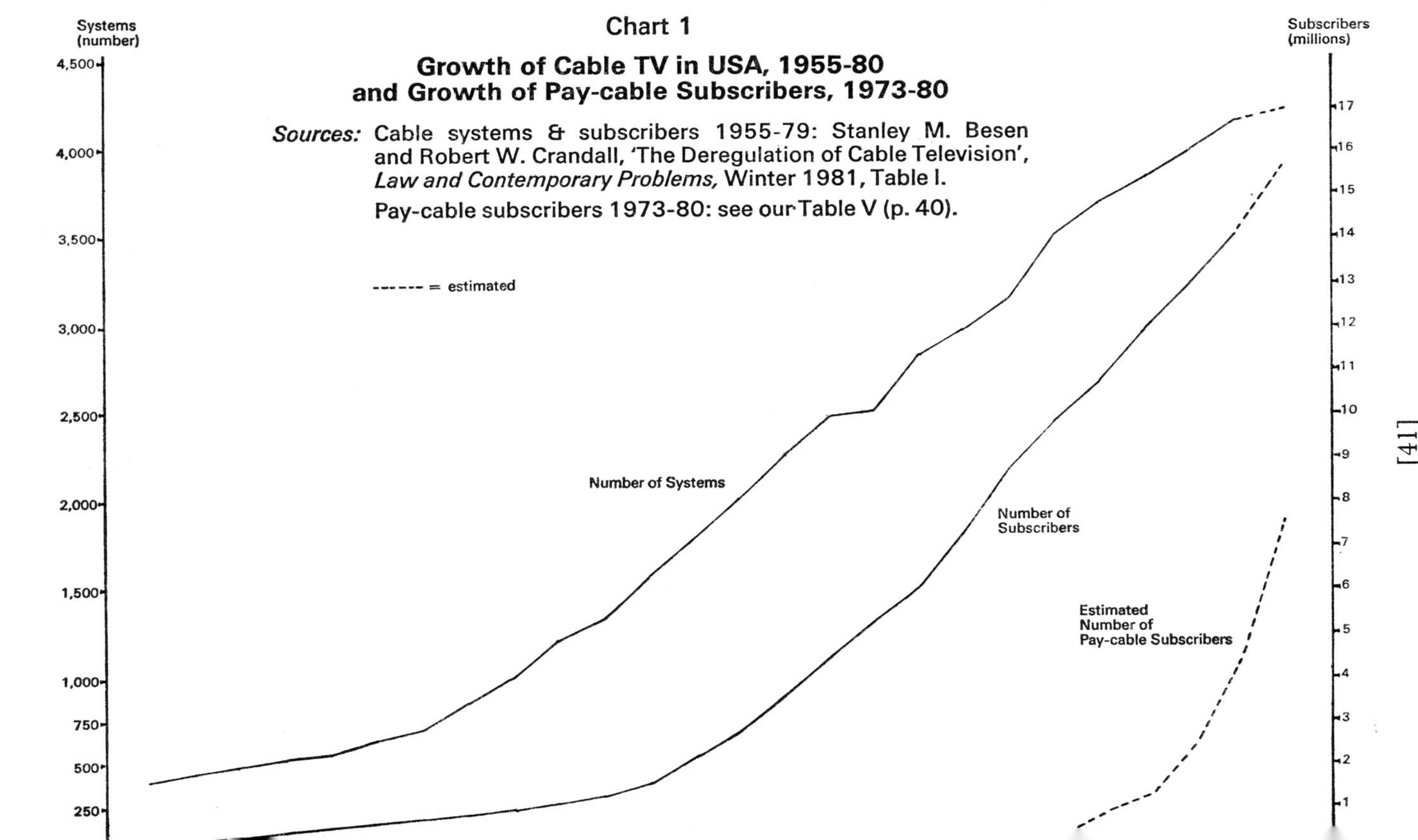

Chart 1

Growth of Cable TV in USA, 1955-80 and Growth of Pay-cable Subscribers, 1973-80

Sources: Cable systems & subscribers 1955-79: Stanley M. Besen and Robert W. Crandall, 'The Deregulation of Cable Television', *Law and Contemporary Problems,* Winter 1981, Table I.

Pay-cable subscribers 1973-80: see our Table V (p. 40).

studio for black-and-white television can be equipped for as little as $30,000 and one for colour television for $200,000. It is estimated that over half the cable systems originate some of their own programmes, amounting to an average of 24 hours of their weekly output.

The average monthly fee for subscribers to the 'basic' service is $8 and the average monthly pay-TV fee is $8.50. Over 2,900 systems (70 per cent of the total) accept advertising on their local origination channels at rates varying from $5 to $300 per minute. Most, however, derive less than 5 per cent of their gross revenue from advertising. Pay-cable is available on 2,000 systems reaching 7·5 million subscribers. Most of the systems report that over 50 per cent of their subscribers receive pay-cable channels (Table V).

THREE: The Market for Television

The idea that television should be provided by commercial interests and be subject to market forces has been rejected until very recently as a plausible policy for Britain. It is widely believed there is something special about television which requires it to be publicly-owned or, at least, publicly-regulated. Those special features are less evident in cable, but many people still think that it too requires significant public regulation. We thus confront a paradox: Why should there be freedom and commercialism in speech and publishing but not in television?

To answer this question requires a proper understanding of how markets in television and communications work or fail to work. This Section introduces the ideas and concepts necessary for such an understanding. A much broader approach is adopted than that of confining the analysis to cable TV —for two reasons.

First, the market for cable TV is not well-defined. It is part of a wider market in communications from which it cannot be divorced and examined separately. That the Hunt Committee was obliged to consider cable TV as if it *were* a separate market was unfortunate and no doubt contributed to the failure of its Report to appreciate the wider economic issues.

Secondly, by extending the discussion to broadcast TV it can be shown that the scarcity of frequencies and their vulnerability to interference pose no peculiar problems for a market, and thus that one of the principal arguments for the regulation of broadcasting is seriously flawed.

1. Consumer Sovereignty

Over two centuries ago Adam Smith asserted that the interests of the consumer were paramount:

> 'Consumption is the sole end and purpose of all production, and the interest of the producer ought to be attended to, only so far as it may be necessary for promoting that of the consumer.'[1]

[1] Adam Smith, *The Wealth of Nations,* Cannan edn., Methuen, London, 1904, p. 159.

The entire edifice of micro-economic theory is based on the simple proposition that the goal of all economic activity is to maximise consumer welfare.

It follows that the central issue in any economic evaluation of the cable industry is the respective contributions of the market and regulation to maximising the welfare of the consumer. If for some reason the consumer is not sovereign in the market-place, either because of monopoly or some impairment of his ability to make informed choices, public policy should be designed to correct it. Likewise, government supervision and controls over industry which do not promote the consumer interest are candidates for reform.

Economics thus equates the 'wider public interest' with the consumer's interest. A particularly apposite formulation is to be found in the Monopolies Commission's report on the distribution of films to cinemas:

> 'In examining whether . . . industry operate(s) against the public interest . . . we judge the effects primarily from the point of view of the consumer; that is to say, we are concerned with the question whether both the production and exhibition of films are effectively geared to meeting the needs of the consumer as regards quantity, quality, choice and convenience.'[1]

This interpretation has been used before and lies at the heart of the Thatcher Government's policy on telecommunications.[2]

2. Economic Efficiency

The goal of maximising consumer welfare is much too vague. For it to be a useful concept it must be possible to measure changes in welfare. And the advantages and disadvantages of alternative courses of action must be valued in some common unit if they are to be compared and balanced against each other. Economists attempt this through the 'measuring rod of money'.

The concept of economic or allocative efficiency permits a quantitative measure of consumer welfare. Stated simply, economic efficiency is that allocation of resources which maxi-

[1] Monopolies Commission, *Supply of Films for Exhibition in Cinemas,* Cmnd. 6667, HMSO, 1966, para. 197.

[2] White Paper, *The Future of Telecommunications in Britain,* Cmnd. 8610, HMSO, 1982.

mises the difference between what the consumer is willing to pay and the opportunity costs of production.

Allocative efficiency is based on a *valuation* of costs and benefits to society as a whole. Thus the benefits to consumers are measured by their willingness to pay for 'quality, quantity, choice and convenience'. The social costs are measured by the value of the output the resources could have produced in their next best use—the notion of social opportunity costs. An efficient allocation maximises the difference between social costs and benefits.

What efficiency is not

This social cost-benefit calculation must be clearly distinguished from several other monetary criteria.

First, maximising economic efficiency is not the same as maximising the market value of goods and services. When a consumer buys an item he is usually willing to pay more than its stated price. The difference between what he is willing to pay and the price he actually pays is referred to by economists as the consumer's surplus. It represents the net benefit the consumer derives from a particular transaction. Similarly, firms receive their 'surplus' in the form of profit. Correctly stated, the welfare goal of economic activity is to maximise the joint surplus of consumers and producers.

Secondly, the social cost-benefit test must be clearly distinguished from industry's profit criterion. For all profitable businesses, the consumer is willing to pay more for their products than it costs to produce them. But this profitability may be achieved at a heavy cost to the rest of the economy if the resources employed are under-valued. If a firm is responsible for considerable pollution, for example, its private costs of production will be lower than social costs. The efficiency calculation would take the third-party losses caused by pollution into account and balance them against the consumer's gain.

Role of efficiency in policy

Although economic efficiency ought to be valued highly, it is only one among many values. Through the government of the day 'society' may, and often does, interpret the public interest in a way which sacrifices economic efficiency for other objectives. This frequently takes place without full information about the costs and the opportunities for economic betterment which are thereby lost.

Fortunately, the present Government's policy on telecommunications generally accords efficiency a very high priority. That priority is affirmed in the clearest possible terms in the recent White Paper on telecommunications:

> 'It is the Government's aim to promote consumer choice. Wherever possible, we want industrial and commercial decisions to be determined by the market and not by the state. We believe that consumer choice and the disciplines of the market lead to more stable prices, improved efficiency and a higher quality of services.'[1]

3. The Market Defined

The notion of a market is frequently defined very narrowly in terms of a particular product or an industry. An economically meaningful definition, however, will embrace not only competition among firms producing a particular product but also competition from other products which consumers regard as close substitutes. Furthermore, the degree of competition among producers and closely substitutable products and services may differ from area to area, giving the market a geographical dimension. The producers of television sets in Australia probably do not compete with those in Sweden because the costs of transport from Australia are high. Australian television makers could compete only if they were able to achieve much lower production costs than Swedish makers.

The product market—substitutability the key

The key element in defining an appropriate product market is the notion of product substitutability. Two products, or services, are part of the same product market if the consumer regards them as close substitutes. British Rail's inter-city services, for example, compete with those of long-distance bus companies for the same customers, some of whom will switch their patronage if the price differential between the

[1] Cmnd. 8610, *op. cit.*, p. 2.

two changes. Train and bus travel are close, but not perfect, substitutes because each offers a different quality of service, such as speed, reliability and comfort.[1] Nonetheless, for the purpose of determining industry performance they can be regarded as part of the same product market.

Cable TV not a well-defined market

Cable TV is not a well-defined product market but part of a wider one which includes other forms of television, entertainment, and many forms of communication. It offers a product similar to broadcast TV, the cinema, video, satellite TV, and even the live theatre. It is also in competition with more 'distant' forms of entertainment such as newspapers, sport and other leisure activities. They constitute the relevant product market in which cable will have to compete for consumers' leisure expenditure. Anyone who doubts that this wider definition of the market is appropriate for cable should pause to consider the impact broadcast TV has had on the cinema. The competition between different methods of delivering filmed entertainment to consumers is vigorous and evidenced today by the inexorable decline of cinema attendances. And it is reflected in the fear that cable will damage ITV and the BBC.

When cable enters the two-way communications field it will be in competition with British Telecom, the Post Office and a host of other marketing and business communication services currently provided by the private and public sectors. The economically relevant product market will then be communications. The capacity of cable to provide a variety of services points to the desirability of an integrated communications policy rather than one fragmented along traditional lines. This broader policy is necessary because all the services will be competing with each other and the competitive forces will affect the performance of each sector of the communications market.

Importance of geographic market

The second aspect of the definition of the market is geographic. Firms selling identical products in two cities located far from

[1] For a theoretical approach which develops the theme of 'product characteristics', Kelvin Lancaster, 'A New Approach to Consumer Theory', *Journal of Political Economy*, Vol. 74, 1966, pp. 132-57.

each other will be in different markets. A cinema in Oxford probably does not compete with a cinema in Banbury, let alone one in Glasgow.

The geographic market is particularly important in the cable industry where, for technical reasons, only one cable system may be financially viable in each area. Although cable is consequently said to enjoy a local monopoly, within each location competition between cable and other services may still be vigorous. This issue will be considered in more detail in Section 5.

4. Forms of Competition

The central economic issue in the current debate on cable expansion is whether a particular mix of competition and regulation brings the maximum benefit to consumers. Competition in the market-place benefits consumers because it ensures that production costs are minimised, that prices accurately reflect those costs, and that producers are responsive to consumers' demand for product variety and quality. A competitive market supplies information and incentives for the efficient allocation of scarce resources to maximise consumer welfare. *The market is a form of regulation: and a competitive market is self-regulating.*

Economists distinguish several forms of competition crucial to an understanding of the cable TV market.

Competition in the market

Competitive pressures *within* a market can take two forms. The first is the competition between closely substitutable products already referred to. The second is competition on the supply side among rival firms. Their jostling for higher profits will ensure that efficient production techniques are adopted and the lowest prices are charged. A firm whose production costs exceed those of its rivals will go bankrupt. A firm which charges more for its product without offering higher quality or something else valued by the consumer will quickly lose all its customers.

Where competition is vigorous, market forces act to discipline those firms which attempt to behave like monopolists or are unresponsive to the demands of consumers. Competitive pressures ensure that firms expand output to the point where price equals marginal opportunity cost. If this condition does not obtain, the implication is that output can be increased at a cost to society which is lower than the consumer's willingness to pay. In a competitive market the profit motive would induce firms to exploit this opportunity for gain, forcing the price down to equate with marginal cost and encouraging cost economies and product innovation.

Potential competition

The performance of an industry is also determined by the threat of competition from outside it. A firm which is able to price its products so as to earn higher-than-normal profits will soon find that others are attracted to enter its industry. Thus the mere threat of entry may be sufficient to induce an industry which might otherwise act monopolistically to price its goods and services competitively. An efficient market may not require many actual sellers so long as there are many potential entrants.[1]

Barriers to entry

The intensity and effectiveness of potential competition depend crucially on the height of the *barriers to entry* in an industry. That term has been defined by Professor George Stigler as

> 'a cost of producing (at some or every rate of output) which must be borne by a firm which seeks to enter an industry but is not borne by firms already in the industry.'[2]

An extreme example of a barrier to entry is a licence to operate a TV station which its holder has obtained from the IBA but which is unavailable to a potential entrant at any price.

The significance of barriers to entry cannot be over-emphas-

[1] The notion of potential competition, or the threat of entry, as a factor influencing an industry's performance was developed at the beginning of the century by J. B. Clark, *The Control of Trusts: An Argument in Favour of Curbing the Power by a Natural Method,* Macmillan, New York, 1901.

[2] George Stigler, *The Organization of Industry,* Irwin, Homewood, Ill., 1968, p. 78; also Joe Bain, *Barriers to New Competition,* Harvard University Press, Cambridge, Mass., 1956.

ised.[1] If they are absent, even markets supplied by a single firm will function competitively. With costless entry, a pricing policy which produces above-normal profits will attract new firms into the industry. The process will continue until all excess profits have been eliminated. Because it can never be sure of its survival with other firms competing for its business, the single firm has an incentive to forestall entry by behaving, from the start, *as if* it were in a competitive market.

The concept of potential competition adds an important, though often neglected, dimension to any analysis of industry performance, requiring full consideration of *prospective* as well as actual market forces. Merely because economies of scale justify only one cable system in each area will not be sufficient to enable the supplier to charge excessive prices. It must be shown, in addition, that barriers to entry in the industry are significant *and* that demand for cable services is relatively inelastic, that is, that the quantity demanded is relatively unresponsive to variations in its price.

5. Broadcast TV without Markets

Broadcasting policy in Britain derives from a view that the market is an unworkable and inefficient means of providing television. That view is shaped by both the technical features of broadcast TV and its capability as an instrument to influence and persuade. Together, these characteristics are claimed to justify extensive public regulation of broadcasting. In particular, the scarcity of frequencies—which limits the number of television channels—and the need to control signal interference are alleged to make a market in broadcast TV unworkable. Both these technical grounds for rejecting the market are, however, seriously flawed.

Critique of technical grounds for broadcast TV regulation

(i) *The scarcity of frequencies*

That frequencies are scarce is not in itself sufficient justification for public regulation of television. All resources are scarce,

[1] Economists are devoting increasing attention to the concept of potential competition: William J. Baumol, 'Contestable Markets: An Uprising in the Theory of Industry Structure', *American Economic Review*, Vol. 72, March 1982, pp. 1-15; William M. Landes and Richard A. Posner, 'Market Power in Antitrust Cases', *Harvard Law Review*, Vol. 94, March 1981, pp. 987-96.

which is why markets are required. Land is scarce, yet it is traded in the market-place every day. The analogy with land is particularly apt since, like the electro-magnetic spectrum, it is in relatively fixed supply. Moreover, that the number of broadcast TV channels in Britain is small has nothing to do with natural or technical factors. It is the outcome of a conscious policy to ration television. The present 'scarcity' of TV channels is the result of, and not the reason for, government policy in the past.

A related argument is that, even were the number of TV channels to be increased, they would still be insufficient to create a truly competitive market in broadcast TV. This argument is considerably overstated for several reasons. First, as emphasised above (p. 49), a market may behave competitively even with only a few firms if there is a real threat of entry. Secondly, the economic model of competition is unclear on how many firms are required to ensure that price equals marginal cost. All that can be claimed with certainty is that an arbitrarily large number of firms is not a necessary condition. Rivalry between a few large firms in a heavily concentrated industry may achieve approximately competitive results provided they do not collude to restrict output.[1]

Finally, even if the limited number of TV channels does give rise to monopoly tendencies and inefficiency, that does not establish the desirability of government operation or regulation. The relevant test should be to compare the outcome of an imperfectly-operating market with the way the regulatory system works. *There is no convincing evidence for the view that an imperfect market is any less efficient than the present system of public service broadcasting.*

(ii) *Controlling signal interference*

The second technical reason why regulation is deemed necessary is to deal with the problem of signal interference. A free market in broadcasting would quickly lead to congestion of the electro-magnetic spectrum—'etheric bedlam produced by

[1] Kwoka's work, for example, suggests that 'industry [profit] margins are unaffected until output control by one or two firms reaches 25 or 30 per cent and that, even then, the problems of co-ordinating three firms are so severe that the presence of a third large firm in the industry can mitigate the effects of such output control'. (John E. Kwoka, Jr., 'The Effect of Market Share Distribution on Industry Performance', *Review of Economics and Statistics,* Vol. 61, February 1979, pp. 101-109, extract from p. 108.)

numerous stations all trying to communicate at once'.[1] Consequently, competition is undesirable.

One obvious solution to this problem is for government to allocate frequencies, restrict entry, and minimise signal interference by imposing appropriate technical standards. Alternatively, broadcasting can be made a monopoly—the course adopted in the 1920s when the BBC was set up.[2] It was thought at the time that a BBC monopoly would be the most efficient way to avoid the sort of signal interference observed in the USA where private broadcasting was permitted. Clearly, a monopolist with control over all broadcasting channels will ensure that signals do not interfere with each other.

This criticism, however, also misperceives the source of the difficulty and its solution. Congestion of the airwaves is caused not by competition but by the absence of clearly-defined and exclusive property rights in frequencies. If no-one has the exclusive right to particular frequencies, users will ignore the costs their transmissions impose on others. The root cause of the congestion difficulty is that, without private ownership of a scarce resource, those who use it will treat it as valueless. It will be wastefully exploited, no-one making an effort to husband it and ensure it is used in the most efficient way.

Markets without property rights

Markets do not work properly unless all resources are owned by somebody.[3] An important function of government is therefore to define the content of ownership rights and to provide mechanisms by which they can be transferred and enforced. Rights to land, for example, are defined by law and government provides courts and police to enforce them and the land registry to help minimise transfer costs. The rest is left to individuals in markets for sale, lease and rental.

A market in broadcast TV would be a workable alternative

[1] Cited in Denis Thomas, *Competition in Radio,* Occasional Paper 5, IEA, 1965 (2nd edn. 1966), p. 10.

[2] Ronald H. Coase, *British Broadcasting: A Study in Monopoly, op. cit.*

[3] On the economic importance of property rights, Armen A. Alchian, *Economic Forces at Work,* Liberty Press, Indianapolis, 1977, Part II; Harold Demsetz, 'Toward a Theory of Property Rights', *American Economic Review,* Vol. 57, May 1967, pp. 347-59; Erik G. Furubotn and Svetozar Pejovich (eds.), *The Economics of Property Rights,* Ballinger, Cambridge, Mass., 1974.

if there were exclusive property rights in frequencies. They would have to specify band-widths and guarantee the owner's right to deliver signals and use his frequency for any purpose not contrary to the general laws of the land. In addition, the owner must be able to transfer and sell his rights at will. This condition is crucial because it ensures that frequencies are put to their most efficient uses by the most efficient users.

Frequency allocation by auction

The idea that frequencies should be allocated by the market mechanism was first put forward by Leo Herzel in 1951. He suggested that, in addition to defining exclusive property rights in frequencies, government should sell them by competitive auction. The Herzel Auction proposal was stated in the following terms:

> 'The . . . [government] could lease channels for a stated period to the highest bidder without making any other judgement of the economic or engineering adequacy of the standards to be used by the applicant. The [government] would still determine the width of the channel but on the basis of one criterion—the maximisation of revenue from leasing this scarce natural resource.'[1]

Herzel's proposal differs radically from the way frequencies are allocated by the Home Office in Britain—namely, free of charge to commercial and public operators. In effect, the Home Office is giving away a valuable resource. This reality is not always appreciated—except, of course, by those commercial undertakings who obtain a licence to broadcast. It was clearly not lost on the late Lord Thomson of Fleet when he referred to his ITV franchise as a 'licence to print money'. The present method of awarding ITV licences bestows large windfall profits on the programme companies,[2] which government has taxed since 1964.[3]

[1] Leo Herzel, 'Public Interest and the Market in Color Television', *University of Chicago Law Review,* Vol. 18, Summer 1951, pp. 802-16.

[2] Brian Hindley, 'The Profits of Advertising-Financed Television', Appendix G to Annan Report (Cmnd. 6753-I).

[3] The programme contractors are required to make two payments to the IBA: a rental for the use of the IBA's transmitters (totalling over £24 million in 1980-81) and what is sometimes known as the 'Exchequer Levy', which is a profits tax. The IBA pays this to the Exchequer. The levy collected in 1980-81 was just over £52 million.

Benefits of auctioning frequencies

Harnessing the price mechanism to allocate frequencies has several attractions.[1] It would be administratively simple; it would not require complex criteria and standards; and it would yield revenue for the public exchequer. Government would merely be required to maximise the revenue from selling frequencies.

The purpose of the Herzel Auction, however, is not primarily to siphon off potential monopoly profits and generate revenue for government.[2] It is to ensure that frequencies are used in the most efficient manner.[3] This condition requires each frequency to be put to the use which maximises allocative efficiency. Pricing frequencies ensures that those who operate on them take into account the economic value of the spectrum. It provides them with an incentive—which the present method does not—to undertake research and introduce new technology to make better use of it.

An administrative agency could assign frequencies to the most efficient uses if it were free of political and bureaucratic pressures and were well informed. But, as Ronald Coase has stressed, administrative agencies suffer from two handicaps compared with the market.[4] First, they lack precise measures of the costs and benefits of each use of the spectrum; and, secondly, they are relatively ill-informed both about the uses of frequencies and the preferences of consumers. Additionally, a government department often has no desire—and little incentive—to promote the efficient use of the spectrum or

[1] For further discussion: Ronald H. Coase, William Meckling and Jora Minasion, 'Problems in Radio Frequency Allocation', Rand Corporation, Santa Monica, 1960, Chap. 4; Arthur S. De Vany, Ross D. Eckert, Charles J. Meyers, Donald J. O'Hara and Robert C. Scott, 'A Property System for Market Allocation of the Electro-Magnetic Spectrum', *Stanford Law Review*, Vol. 21, June 1969, pp. 1,499-1,561.

[2] Although the Exchequer Levy can be viewed as a 'price' for the right to use frequencies, it does not encourage the efficient allocation of frequencies. It is simply a profits tax which does not indicate to the programme companies the opportunity costs of their use of frequencies.

[3] Despite the evident attractions of the auction method, it has met with a cool reception. The US Federal Communications Commission (FCC) recently ruled that the auction of seven DBS transponder leases, which were sold for US $90 million, was invalid. Interestingly, the FCC later allowed the lessor (RCA) to lease each transponder at the average price.

[4] Ronald H. Coase, 'The Federal Communications Commission', *Journal of Law and Economics*, Vol. 2, October 1959, pp. 1-40.

respect the wishes of the consumer. This criticism is amply vindicated by the recent decision of the Home Office obliging the cable relay service in Wales to take the Welsh-speaking version of the new Channel 4 despite the overwhelming vote by subscribers in Wales for English-language programmes.[1]

Criticism of a price mechanism

The proposal that broadcast TV should be governed by market forces has been heavily criticised as impractical. Apart from a concern about programme standards and monopoly, which will be discussed later, two difficulties have been identified.[2]

(i) *Rights cannot be clearly defined*

First, it is claimed that, even if the rights to a particular frequency were exclusive, it would not make the market workable. It would certainly prevent others from using the frequency but it would not eliminate interference from adjoining frequencies. Government regulation would still be required to minimise interference. Hence, the net result would be merely to superimpose payment for operating licences on the present regulatory system.

Again, this claim is ill-founded; the possibility of incompatible uses of adjoining frequencies poses no difficulties for the market. To understand why, let us consider the case of a TV company whose transmission is not being received clearly by viewers because of the activities of another user of the spectrum. Let us assume that, although each owner has the exclusive right to a particular frequency, the law is silent about liability for interference. It would then be in the interests of the aggrieved party to negotiate with the party causing the interference about steps to reduce it.

In principle, a free market in frequencies could find a solution through negotiation.[3] The aggrieved party would be willing to pay for a reduction in interference a maximum sum equal to the amount his profits are increased by preventive

[1] 'Cable Votes', *Economist*, 30 October 1982, p. 33.

[2] Dallas Smythe, 'Facing the Facts about the Broadcast Business', *University of Chicago Law Review*, Vol. 20, Autumn 1952, pp. 96-106; and, more recently, William H. Melody, 'Radio Spectrum Allocation: The Role of the Market', *American Economic Review*, Vol. 70, May 1980, pp. 393-97.

[3] Ronald H. Coase, 'The Problem of Social Cost', *Journal of Law and Economics*, Vol. 3, October 1960, pp. 1-44.

measures. Similarly, the party causing the interference would be willing to take abatement measures so long as their cost was lower than the sum the aggrieved party was willing to pay. In short, so long as the payments exceeded the abatement costs, the parties would continue to bargain for a reduction in interference. The solution finally negotiated would be efficient in the sense of maximising the joint value of the output obtainable from both frequencies.

This market solution would, however, differ from the present goal of spectrum management which is to minimise signal interference. This goal ignores both the direct costs of abatement devices and the indirect costs of lost output incurred in reducing interference. The economically efficient degree of interference would balance these costs with the benefits, and would necessarily be higher than that technically possible. To put the same proposition another way, reducing signal interference to its technically feasible minimum imposes an unnecessary and wasteful cost on society.

(ii) *Non-commercial use of the spectrum*

Much of the broadcasting spectrum is used by non-commercial interests such as the armed forces, police, and government departments, frequently to provide essential services. It is argued that such non-profit-making organisations could not seriously be expected to pay for access to the spectrum— and that it would, moreover, entail no more than a pointless transfer of taxpayers' money from one government department to another.

It is, however, by no means clear why such users should not pay for frequencies. As Herzel notes, they 'compete for all other equipment or else they don't get it'. To repeat, the main purpose of pricing frequencies is not to generate revenue but to ensure that all users take into account the opportunity cost of their activities. There is no reason why non-commercial users should remain unaware of this cost and it would seem a prerequisite for rational planning of spectrum use that they should. If government departments and other public bodies had to pay for access to frequencies, the cost would be definite and transparent. The Treasury could then decide at budget time whether each use was as important as its cost, thus ensuring higher efficiency in government.

The problems caused by the free allocation of frequencies are well illustrated by citizens' band (CB) radio. A valuable resource was recently allocated to this use with hardly anyone being aware that a cost (in the sense of an alternative foregone) was incurred. It is no wonder that users of the spectrum constantly clamour for more space when they have to pay nothing for it! Not least among the benefits of a market for frequencies would be the injection of an element of rationality into public debate on topics like CB radio where there is currently only muddle.

Source of market failure

The preceding analysis argues that neither scarcity of the airwaves nor signal interference provides a rational basis for public regulation of broadcasting. If there is a case for the present arrangements, it must be because a market in frequencies would not bring about the efficient use of the spectrum. Although a number of reasons are advanced to suggest why this might be so, they all boil down to one—the high costs of operating in that market. The costs of negotiating, monitoring and enforcing bargains about frequencies—known to economists as 'transactions costs'—may be so large in practice as to prevent the market working properly or even at all. It almost goes without saying that this proposition has never been convincingly established. In practice, as we have argued, a more market-oriented scheme for allocating frequencies with some very attractive features has simply been dismissed by casual reference to a few potential deficiencies. Regulation also has deficiencies and can be justified only if it secures economic benefits in excess of its costs, both direct and indirect in the form of inevitable misallocations elsewhere in the economy.

6. Implications for Policy

Although the above discussion may have seemed somewhat tangential to the controversy surrounding the de-regulation of cable TV, the issues it raises are at the crux of the matter. We have argued that the main reasons advanced for excluding the market from broadcast TV are deeply flawed; and that a market in broadcast TV is feasible and could handle the technical problems which have been held to justify the present public broadcasting system. There are strong grounds for supposing that a market in frequencies, augmented by public regulation to deal with some real deficiencies which would exist, would be more efficient than the present arrangements. The hidden disadvantages of British broadcast TV in terms of foregone opportunities, unnecessarily high costs and monopoly profits are large because of a deliberate policy to ration television and place it in the hands of one public monopoly and a number of regional private ones.

The lesson of this Section is the importance of getting the principles governing policy on cable TV right from the beginning. For, as Ronald Coase has written:

> 'It was indeed in the shadows cast by a mysterious technology that our views on broadcasting policy were formed . . . The problems posed by the broadcasting industry do not call for any fundamental changes in the legal and economic arrangements that serve other industries. But the belief that broadcasting is unique and requires regulation of a kind which would be unthinkable in the other media of communications is now so firmly held as perhaps to be beyond the reach of critical examination. The history of regulation in the broadcasting industry demonstrates the crucial importance of events in the early days of a new development in determining long-run governmental policy.'[1]

[1] Ronald H. Coase, 'The Federal Communications Commission', *op. cit.*, p. 40.

FOUR: Financing and Programme Standards

A market in television cannot be analysed in the same way as that in bread or bolts. It has two special features which may give rise to difficulties: the subtle effects of financing on programme standards, and the so-called local monopoly problem. In this Section we consider the former in some detail and argue that pay-TV is a superior method of financing television.

Financing and programme standards are at the centre of the current debate about cable TV. What has not been properly appreciated is that the *source* of finance has crucial effects on programme standards. Many claims have been made during several months of public debate about the impact of unrestricted cable TV on programme standards. It has been contended that cable TV will fragment audiences and cream off the best programmes, thereby impoverishing broadcasting; or, alternatively, that the logic of profit dictates it will produce only low-quality programmes. One mildly unfortunate consequence is that this Section must be, to some extent, a catalogue of such fallacies and an analysis of the errors which gave rise to them.

1. Methods of Financing TV

There are three principal ways of financing television: taxation (the TV licence fee), advertising, and a direct charge to the viewer. The Annan Committee enunciated the principle that the different forms of broadcasting should not compete for the same source of finance.[1] Several submissions to the Hunt Committee advanced the same argument, proposing that cable TV should be financed by subscription and not by advertising. The Hunt Committee, however, rejected this approach and recommended that cable operators be permitted to obtain revenue from both sources.[2]

We will briefly consider the effects of each of the three

[1] Cmnd. 6753, *op. cit.*, para. 7.6.

[2] Paras. 31, 49.

methods of financing television on the standard and variety of programmes.

(i) *The licence fee*

The BBC is financed by a yearly lump-sum tax (the 'licence fee') on each household with one or more TV sets. The tax is currently £15.00 for monochrome and £46.00 for colour.[1] The BBC is mandated by its Charter to inform, educate and entertain. In undertaking this task, it must maintain high standards, balance and impartiality in its programmes and not offend good taste and decency.

The revenue from the licence fee provides the BBC with an indication of its success in appealing to a mass audience. But it is a very crude guide which merely reflects the viewer's all-or-nothing decision to have a TV set or not. The link between the *intensity* of viewers' preferences and the programmes provided by the BBC is thus very imperfect—indeed, very nearly non-existent. Moreover, since the BBC has to compete with ITV for the same mass audience, it is obliged to offer a programme schedule not dissimilar from ITV.[2]

(ii) *Advertising*

Broadcast and cable TV financed wholly by advertising have inherent biases towards inefficiency in programming because neither caters directly to the consumer's demand. Rather, they sell audiences to advertisers. Thus the character of programmes shown on advertiser-supported TV is determined by the usefulness of the programme in selling the advertiser's product. The income from advertising of the TV companies depends more or less on the size of the audience for each advertisement. The mass audience is crucial to the profitability of the ITV companies; thus they have a tendency to offer

[1] Although the licence fee is levied only on those who own or hire TV sets, the revenue it produces is used to finance BBC radio also. The revenue is distributed between TV and radio in the proportion of 70:30.

[2] As William Cotton, the BBC's Director of Programmes, has said: 'As a duopoly, we only compete for the attention of our audience, not for our income'. (Milne, *op. cit.*, p. 19.) Before the advent of Channel 4, the total television audience was divided equally between BBC and ITV, with 42 per cent viewing BBC1, 8 per cent BBC2, and 50 per cent ITV.

programmes which nearly everyone is *willing* to watch, even though no-one wants very much to watch them.[1]

It will be merely by *chance* that advertiser-financed TV will broadcast the type and range of programmes consumers value most highly, since the latter have only a crude and indirect way of expressing their preferences.[2]

This deficiency, however, does not necessarily imply that financing TV in this way is undesirable, for TV advertising serves the important economic function of disseminating information about products. The programme companies supply two 'products'—audiences to advertisers and entertainment to viewers—but *sell* only to the advertisers. Since the viewer receives his entertainment (the programmes) free, a system of advertiser-financing generates a huge consumers' surplus. It is those people who buy the advertised products who pay and, to the extent that they differ from viewers, a significant redistribution of wealth takes place. TV financed by advertising, therefore, is not truly a market in television but a subsidiary activity of the advertising industry. What *is* undesirable is that the commercial sector of a country's TV system should be financed *solely* by advertising. Pay-TV is quite different: it permits a market devoted to entertainment in which the preferences of viewers count.

(iii) *Pay-TV*[3]

Of the three methods of financing, pay-TV is the only one that provides TV programmers with a direct measure of the preferences of viewers. It has clear attractions from an economic point of view. Subject to an important reservation discussed

[1] The best discussion of the literature on advertiser-financed TV is in Bruce M. Owen, Jack H. Beebe and Willard G. Manning, Jr., *Television Economics,* D. C. Heath, Lexington, Mass., 1974, Chap. 3.

[2] The IBA's programme regulations require the ITV companies to show minority, educational and public service programmes, thus mitigating the tendency towards mass-appeal programmes. The programmes which result from the IBA's intervention, however, satisfy not consumer preferences but the IBA's 'programme tastes'—and therefore do not necessarily offer what consumers want to view.

[3] For discussions of pay-TV, David Sawers, 'The Sky's the Limit', in Wilfred Altman *et al., TV: From Monopoly to Competition—and Back?,* Hobart Paper 15, IEA, Rev. Edn., July 1962, Part III; Sir Sydney Caine, *Paying for TV?,* Hobart Paper 43, IEA, 1968; Roger G. Noll, Merton J. Peck and John J. McGowan, *Economic Aspects of Television Regulation,* Brookings Institution, Washington DC, 1973, Chap. 5.

later (p. 63), the present system of zero-priced TV at the point of consumption is incompatible with the efficient use of resources in programme production and distribution since it does not register, nor is it responsive to, the intensity of viewers' preferences for programme type and quality.

In contrast, pay-TV can survive only if it provides the consumer with a package of programmes he chooses to continue to pay for. The more intense a consumer's preference for a type of programme, the more he will be willing to pay. Profit-maximising cable programmers will thus offer a mix of channels and programmes which caters for market demand and diversity. As Professor Jora Minasian has observed:

> '. . . a subscription system can be expected to yield a more diversified program menu than an advertising system because the former enables individuals, by concentrating their dollar votes, to overcome the "unpopularity" of their tastes'.[1]

Key requirements: unlimited channels and direct payment

Unlimited channels and competition ensure that all groups with intense preferences are catered for *provided* their willingness to pay is sufficient to make it profitable to produce and distribute a programme. Unlimited channels and direct payment by viewers are *both* crucial. Without the former many minority audiences would remain excluded. Without the latter many programmes, especially those which satisfy intense preferences, would not be made since willingness to pay could not be registered in the market for programme production. This does not mean that pay-cable will not screen mass-appeal programmes; their large audiences will make them relatively profitable even at low prices. The case for pay-TV rests, not on its ability to offer minority-taste programmes, as is frequently argued, but on its responsiveness to consumer preferences.

If all channels are controlled by one company, the efficiency of the pay-cable market is somewhat impaired. Although a monopolist will still have an incentive to produce the most

[1] Jora R. Minasian, 'Television Pricing and the Theory of Public Goods', *Journal of Law & Economics,* Vol. 7, October 1964, pp. 71-83, extract from p. 75. There will, of course, always remain some minorities too small to be served. In the Hartford subscription-TV experiment in Connecticut, for example, the least watched programme, viewed by one household, was 'You and the Economy' featuring a panel of Yale economists. (Noll, Peck and McGowan, *op. cit.*, p. 133.)

preferred programmes (which would not be true if his revenue came from advertising), he is likely to produce fewer programmes and charge subscribers higher prices than under competition. The reason is that he will take into account the effect on his total profits of diverting audiences from one of his channels to another. Thus, where audience diversion is significant, the *net* profitability of providing an additional channel will be lower to a monopolist than to a programme supplier subject to competition. In catering to the preferences of viewers, however, a monopoly financed by subscription is preferable to one financed by advertising.

2. Cable TV as a Public Good

Television broadcasts have a peculiar characteristic which complicates the analysis. A broadcast programme can be viewed by all who have a set. This quality distinguishes it from most other goods and services—a particular loaf of bread, for example—which, if they are consumed by one individual, cannot be consumed by any other. Television is what economists call a public good[1] because one person's viewing of a programme or channel does not diminish the amount of viewing available to anyone else.

The joint-consumption characteristic of cable TV programmes gives rise to a real inefficiency. Once a programme has been produced and shown to some subscribers, it can be screened to additional viewers at no extra cost. Since the marginal cost of serving an additional viewer is zero, some people have concluded that pay-TV is inherently inefficient. This conclusion is incorrect since the inefficiency arises not from charging for TV but from the necessity of charging each viewer the same price, thereby excluding those who are willing to pay less to see a programme.

Ideally, consumers should be charged different prices so that no-one who places some positive money value on a programme is excluded. In practice, price discrimination is not feasible since information about individual viewers' preferences would be inordinately costly to obtain. Many, if not all, viewers

[1] The classic discussion of a public good can be found in Paul A. Samuelson, 'The Pure Theory of Public Expenditure', *Review of Economics & Statistics*, Vol. 36, November 1954, pp. 387-89. Also John G. Head, *Public Goods and Public Welfare*, Duke University Press, Durham, 1974, Chap. 3.

would have an incentive to claim to value a programme less than they actually did in order to obtain it at a lower price. Since the cable operator would not be able to distinguish the truthful from the untruthful, he is obliged to sell each channel/programme at a uniform price to all viewers.

The cable operator can, however, distinguish and discriminate between *groups* of consumers with different preferences. The virtually unlimited channels of cable TV will enable him to charge a different price for each channel and thereby practise a crude form of group price discrimination (considered further in Section 5).

Thus, while the revenue of cable operators will provide only a crude measure of the intensity of preferences, and while pricing TV will inefficiently exclude some viewers, pay-TV does have a distinct advantage over broadcast TV where there can be no direct link at all between consumer demand and the supply of programmes.

3. On Giving People What They Want

Behind much of the concern about broadcasting standards is an élitist view of what the public should be offered. This approach owes much to the influence of Lord Reith, the first Director General of the BBC. To the criticism that the then BBC monopoly failed to offer programmes people wanted, he replied in characteristic manner:

> 'It is occasionally indicated to us that we are apparently setting out to give the public what we think they need—and not what they want, but few know what they want and very few what they need.'[1]

This lack of faith in the consumer must be repugnant to a free society. The idea that, in a democracy, a body controlled and operated by government should have a monopoly over so important a form of communication is extraordinary. As Professor Peter Wiles has written:

> 'To have maintained democracy despite such an institution [i.e., the BBC] is a *tour de force* of which Britain has every reason to be proud. But even a tight-rope walker is safer on the ground.'[2]

[1] Cited in W. Altman *et al., op. cit.,* p. 13.

[2] P. Wiles, 'Pilkington and the Theory of Value', *Economic Journal,* Vol. LXXIII, June 1963, pp. 185-200, extract from p. 197.

There is good reason to believe that, because of the regulated structure of broadcasting, programmes have tended to reflect middle-class values and tastes. Arguments about maintaining standards often boil down to preserving a type of programme which satisfies influential sections of society who fear those presently not catered for have baser tastes. Perhaps the most cogent rebuff to the critics of pay-TV was expressed more than 20 years ago by Sir Robert Fraser, the first Director General of the Independent Television Authority (ITA):

> 'If you decide to have a system of people's television, then people's television you must expect it to be. It will reflect their likes and dislikes, their tastes and aversions, what they can comprehend and what is beyond them. Every person of common sense knows that people of superior mental constitution are bound to find much of television intellectually beneath them. If such innately fortunate people cannot realise this gently and considerately and with good manners, if in their hearts they despise popular pleasures and interests, then of course they will be angrily dissatisfied with television. But it is not really television with which they are dissatisfied. It is with people.'[1]

4. Clarifying the Issues

The debate over cable TV has been marred by a failure to distinguish clearly pay-cable from advertiser-supported TV and to state precisely how good programme standards are to be defined. This confusion has led to several misleading criticisms of cable TV.

The American network experience

It has been claimed that British television is 'better' than North American, that what keeps ITV programmes from falling to American network standards is public regulation, and that unregulated cable TV would wreak havoc with the high standards of British television.

The first assertion is difficult to rebut, not because it is necessarily true but because evaluating the differences between British and American (or any other) television is extremely subjective. Hence anyone can hold strong views on the matter

[1] Speech delivered to the Manchester Luncheon Club, 17 May 1960, cited in Denis Thomas, 'Commercial TV—and After', in Altman *et al.*, *op. cit.*, pp. 64-5.

secure in the knowledge that his opinions cannot be refuted by facts. The authors believe that British television does have more programmes of the sort educated people like, as is to be expected since such people control the IBA and the BBC. Whether it satisfies the wants of the less intellectual and less well-educated is more doubtful.

It is nonetheless true that unregulated advertiser-supported TV has a tendency to screen mass-appeal programmes which are aimed at the lowest common denominator. It is the *source* of the US networks' revenue which gives rise to this problem and not their private ownership. As has been seen, the economic incentives for privately-operated *pay*-cable TV are entirely different. Pay-TV with unlimited channels has a tendency to diversify rather than offer lowest common-denominator programmes.

An unedifying aspect of the recent debate surrounding cable has been the attempt by broadcasting interests to persuade the public that cable TV will not enrich choice but, in the words of the BBC, simply 'coarsen popular tastes'.[1] Evidence to the contrary is provided in Table VI which contrasts the television programmes available to viewers in New York and London respectively at a random time and date in mid-1982.

Are high standards incompatible with profit seeking?

Quality is a subjective notion. Despite differing opinions about what constitutes good quality, there is no reason why commercial pay-TV and high programme standards should be incompatible. A profit-maximising firm in a competitive environment will supply what the market wants, taking into account the views of *all* consumers. Cable TV will supply more television and hence more variation in programme standards; commercial pay-TV will tend towards neither low-quality, nor high-quality, nor mass-appeal programmes, but towards the most profitable degree of variation in quality.

Economic theory enables us to give the concept of high quality and good programme standards an operational definition. Programmes of the standard to which British viewers are accustomed are costly to produce and are made at the expense of both more programmes and more variety. Other things being equal, the quality of a programme varies with

[1] A. Milne *et al.*, *op. cit.*, Appendix, p. II.

TABLE VI

TV VIEWERS' CHOICE IN NEW YORK AND LONDON AT 9.00 p.m. ON 7 JUNE 1982

Manhattan Cable	BBC/ITV
1. *MASH*	1. News
2. *Black Ghetto Life* (documentary)	2. *Hitch-hiker's Guide to the Galaxy*
3. *Sister, Sister* (film)	3. *Minder*
4. *Merv Griffin* (talk show)	
5. *The Kennedy Years* (documentary)	
6. Baseball	
7. Spanish Play	
8. Variety Show	
9. *Adam and Eve,* with Nureyev (dance)	
10. *Attack* (film)	
11. Spanish Drama	
12. *Orpheus* (opera)	
13. International Education (public access discussion)	
14. Seminar on Nuclear Arms	
15. Baseball	
16. *Bye, Bye Birdie* (film)	
17. *Danger UXB* (drama)	
18. *Dog Day Afternoon* (film)	
19. Gymnastics	
20. Classified Advertisements	
21. Royal Ballet	
22. Folk Art (discussion)	
23. Chinese Cooking	
24. News	
25. *High Country* (film)	

Source: Andrew Neil (ed.), *The Cable Revolution—Britain on the Brink of the Information Society,* Visionhire Cable, London, 1982.

the expenditure on it. More money will hire better actors and directors, will pay for more expensive costumes and sets, and will allow more rehearsal, re-shooting and re-editing.

A priori, there is no upper limit to the amount that could or should be spent, nor any technically-determined minimum either. In general, viewers of the programme will place a higher value on higher quality. More expenditure and hence higher quality are justified so long as the value to viewers of the extra quality at least equals the cost of supplying it.

The cost-justified quality standard

This is the cost-justified quality standard. It implies that, in the same way that programme standards can be too low, they can also be too high. It would not make sense to produce all programmes to the highest technical or artistic standards because their cost would exceed the value to viewers.

To take an everyday example, no-one would seriously argue that all cars should be built to the specifications of a Rolls-Royce. If they were, transport by car would be unnecessarily expensive and enjoyed only by a wealthy few. It would be equally absurd to require, say, a local community or ethnic minority programme produced by a group of concerned citizens to conform to the standards of production of the recent ITV serialisation of *Brideshead Revisited.*

The high channel capacity of cable permits low- (and high-) cost programmes, provided they are what paying viewers want. Perhaps many will be judged to be technically inferior by today's standards. But as long as the objective is to satisfy consumer demand, such a change in programming standards will be desirable. The trade-off between quality and quantity was eloquently recognised by Sir Philip Goodhart, MP, in a recent Commons debate on cable:

> 'Cable television, like any changing communication, will produce as many bad programmes as good ones. There is always the risk that expansion will produce some lowering of standards. It is worth remembering that in civilisation's long history the greatest single decline in cultural, aesthetic and philosophical standards was produced by the printing press. Until then, monasteries were major suppliers of books. The monks diligently produced bibles, prayer books, hymnals and works of theology and philosophy. The printing press came and lesser works were immediately available to the public and it became possible to publish works of casual playwrights like William Shakespeare—a decline perhaps from the bible.'[1]

[1] *Hansard,* Vol. 23, 2 December 1982, cols. 469-70.

Will cable TV be 'more of the same'?

Two observations can be made about the criticism that cable TV will lead not to more choice but simply to more of the same and to endless repeats of old programmes.

First, if cable is to compete with broadcast TV, it must offer either newer and superior programmes or similar ones at more convenient times. As one submission to the Hunt Inquiry stated:

> 'Cable television will not grow in the United Kingdom unless people are prepared to pay for it, and they will not pay for it unless it provides them with services that they value.'[1]

That is the logic of the market-place which many of the critics of cable TV have failed to grasp. Cable can survive only if it meets the unsatisfied demand left by broadcast TV.[2]

Secondly, in one sense the criticism has a certain validity—but only in the short run. The increase in the demand for television programmes stimulated by cable expansion will be hard to satisfy immediately. Programme standards may indeed fall—especially if the demand for cable is large—since the existing talent will have to be spread more thinly. But this should be a transitional problem. If the demand for TV programmes increases, more people will eventually be drawn into the business of acting and producing programmes, and more competition for talent will have the additional benefit of giving actors and producers a freer market in which to sell their skills.

To summarise, higher programme standards are not always preferable to lower. The optimal standard for television, like all products, is one that balances costs and benefits. The production and standards of programmes will be determined by the forces of supply and demand. If the demand for cable TV proves to be large, the profit motive will ensure that more programmes of the type desired by viewers are produced. If

[1] *Charterhouse Japhet Submission to the Hunt Inquiry,* May 1982, p. 1.

[2] There is clear evidence of unsatisfied demand in the home entertainment market. The National Consumer Council's survey of consumer views on leisure found that lack of choice in television programmes and channels was the largest single source of dissatisfaction. Of the respondents, 41 per cent said they were dissatisfied, although only 16 per cent of these considered lack of choice a 'serious problem'. (National Consumer Council, *An Introduction to the Findings of the Consumer Concerns Survey,* 1981, p. 132.) Another indication is the explosive growth of the video market in Britain in the last two years.

cable operators do not respond in this way they will lose their customers and eventually their profits. That is the overriding logic of the market-place.

5. Conclusions

There are sound theoretical reasons for believing that pay-TV is the best practicable method of financing television. The alternative methods—taxation and advertising—are inherently incapable of satisfying the preferences of viewers to anything like the same degree. In particular, they are beset by problems of duplication, of lowest common-denominator programming, and of programming to suit the tastes of the regulators rather than the viewers. The strident and often scornful attacks on popular TV—and on cable TV in particular—are at bottom an expression of contempt for the tastes of viewers in the light of irrelevant comparisons with advertiser-supported TV in the USA.

FIVE: The Local Monopoly Problem

A further peculiar feature of the cable industry is the so-called local monopoly problem. It is claimed that the large investment required to lay cable will result in a single cable system for each geographical area. This problem is recognised by Hunt in the following terms:

> 'The cable operator will have an effective local monopoly as a result of the high cost of installing and operating a cable system and of the need to achieve high customer penetration for financial viability.'[1]

This factor constitutes the sole basis for Hunt's recommendation that cable operators should be franchised. The Report states that such a local monopoly

> 'should be conferred only after an opportunity for judging any competing bids and for securing the provision of the best service for the area concerned'.[2]

Unfortunately, the Report neither clearly defines nor analyses what it means by 'an effective local monopoly'. In this Section we examine critically the issues raised by the geographical monopoly that the owners of cable systems may have.

1. Monopoly

A monopoly exists where the sole seller of a product has the ability to raise its price by limiting production without attracting competition.[3] The monopolist diminishes consumer welfare by creating an artificial scarcity in order to charge a price and earn a profit in excess of what could be maintained under competitive conditions. The social loss from monopoly is measured not by the excess profit but by the reduction in the

[1] Para. 18. [2] Para. 85.

[3] A more detailed discussion is in F. M. Scherer, *Industrial Market Structure and Economic Performance*, Rand McNally, Chicago, 2nd edn., 1980, Chap. 2; and Charles K. Rowley, *Antitrust and Economic Efficiency*, Macmillan, London, 1973.

consumer's surplus at the lower output necessary to enforce the monopoly price.

As used by economists, the term 'monopoly' does not necessarily denote the supply of a product or service by a single firm. A single firm is a monopolist only if it has *market power.* And it will have market power only if it is shielded from potential competition by significant barriers to entry *and* the demand for its product is relatively unresponsive to changes in price.

The latter condition is crucial. When demand is relatively insensitive to price increases,[1] the revenue lost as a result of some reduction in sales will be less than the revenue gained from the higher price of the units still sold. Conversely, when demand is relatively sensitive, a price increase will lead to a net reduction in total revenue. In the limiting case of perfectly elastic demand, any price increase will result in no sales at all. The ability of a single seller to charge a price above marginal cost is then non-existent.

2. Natural Monopoly

A less familiar concept is that of a *natural* monopoly, which can be defined as existing if the entire (actual and potential) market can be supplied at the lowest cost by one firm.[2] A sufficient condition for a natural monopoly is that unit or average costs decrease over the whole range of output necessary to supply the market.[3] Where such economies of scale[4]

[1] In technical terms, when the price elasticity of demand is less than unity.

[2] The concept was first introduced into economics by John Stuart Mill and refined by the American institutionalist economist Richard Ely: J. S. Mill, *Principles of Political Economy*, Longmans, Green, London, 1909, p. 143; Richard T. Ely, *Outlines of Economics*, Macmillan, New York, 1937, p. 628.

[3] Alfred Kahn defines a natural monopoly thus: 'A critical and—if properly defined—all-embracing characteristic of a natural monopoly is an inherent tendency to decreasing unit costs over the entire range of the market'. (Alfred E. Kahn, *The Economics of Regulation: Principles and Institutions*, Vol. 2, Wiley and Sons, New York, 1971, p. 119.) For Stephen Littlechild the pre-condition is 'an industry in which economies of scale are so great, compared to the size of the market, that it is inefficient to have more than one firm producing the industry's output, and in fact, only one firm would be able to survive in such an industry'. (Stephen C. Littlechild, *Elements of Telecommunications Economics*, Peter Peregrinus for the Institution of Electrical Engineers, Stevenage, Herts., 1979, p. 199.)

[4] Economies of scale are obtained when an expansion of x per cent in the quantity of all inputs leads to an increase in output of more than x per cent. This definition assumes that technology is fixed and that there is no excess capacity.

obtain, one producer—the financially strongest or the best-managed—will drive out the competition. The monopoly is natural in the sense that a single firm is required to minimise production costs.[1]

In his treatise on the economics of regulation, Alfred Kahn emphasises that the principal source of decreasing unit costs is 'the necessity of making a large fixed investment in order to serve the market'.[2] The laying of railway tracks and water and gas pipelines and the generation of electricity fall into this category. Such large fixed costs imply that average costs can be reduced—often dramatically—by supplying additional units or serving more customers. Kahn suggests that unit costs will have a stronger tendency to fall if an industry has the following characteristics: the nexus between the supplier and the customer or locality is essentially immovable; the service is non-storable; supply has to be instantaneous; and there are wide fluctuations in demand. All but the first of these factors necessitate heavy capital investment in capacity to meet peak demands.

Cable seems to possess all these properties except the last. Certainly, laying the cable involves a big investment in fixed capital. But the service cable supplies—at least, when it comes to television—is not subject to the type of wide fluctuations requiring extra or excess capacity. Peaks in demand for programmes can be met at no additional cost.

3. Is Cable a Natural Monopoly?

Whether an industry is a natural monopoly can be determined only by careful empirical investigation. There is, unfortunately, very little firm evidence on cable.

An early study by Comanor and Mitchell[3] found that the requirement of the US Federal Communications Commission (FCC) that a cable system with 12 channels must expand its capacity to 20 increased capital expenditure by about 20 per

[1] For a more rigorous definition of a natural monopoly, William J. Baumol, 'On the Proper Cost Tests for a Natural Monopoly in a Multiproduct Industry', *American Economic Review*, Vol. 67, December 1977, pp. 809-22.

[2] Kahn, *op. cit.*, pp. 119-23.

[3] William S. Comanor and B. M. Mitchell, 'Cable Television and the Impact of Regulation', *Bell Journal of Economics and Management Science*, Vol. 2, Spring 1971, pp. 154-212.

cent. That a 67 per cent expansion of channel capacity was achieved with only a 20 per cent increase in capital expenditure suggests that scale economies are significant.

Eli Noam has carried out the most rigorous empirical analysis using data for 1980 and covering nearly all cable systems in the USA (over 4,200).[1] His results show significant economies of scale when output is defined in terms of the number of subscribers but not when defined in terms of the number of homes passed. He concludes that any economies which exist arise not from the technical aspects of cable laying but from definitions of output (such as revenue and the number of subscribers) which include 'a strong element of marketing success'. Noam's study further indicates that experience in operating a cable system (as measured by the number of years of operation) also reduces average costs. Thus decreasing unit costs in cable stem not from the physical plant but from the ability of the largest operator to package and sell his services more efficiently.

The American experience also suggests that cable owners will have a local monopoly. Whilst most US cable franchises are initially awarded to one firm, their terms do not exclude others. Yet only 10 out of 4,200 franchise areas have more than one cable system (called 'overbuild' by the industry), and there is some indication that their subscription rates are higher than the national average.

4. Efficient Pricing

Natural monopolies give rise to another problem. For there to be allocative efficiency in an economy, firms must set prices equal to marginal cost. A natural monopoly, however, would make losses if it adhered to this socially optimal condition. Once the cable system has been built, the marginal cost of connecting one more subscriber is low. If the cable operator charged each subscriber only the connection and related service costs he would obviously not even begin to cover his large fixed costs. Because a natural monopoly has high fixed but low variable costs it cannot operate profitably if it supplies its service at a price equal to the latter.[2]

[1] Eli M. Noam, 'Economies of Scale in Cable Television', Columbia University, New York, August 1982 (unpublished).

[2] The reader will recognise that this problem is formally identical to that discussed in relation to the public-good nature of television (Section 4, pp. 63-64).

To stay in business the owner of a cable system might be forced to adopt a pricing method that leads to social inefficiency. He could charge each subscriber a price which covered a proportionate share of the fixed costs plus a reasonable rate of return on capital. This method would be inefficient because it would discourage potential consumers who were willing to pay at least the direct costs of supplying them with cable TV.

5. Price Discrimination—a Solution

There are two ways of dealing with the problem posed by ever-decreasing average costs.[1] The first is for government to subsidise the monopolist to enable him to price at marginal cost. Unfortunately, he then has no incentive to produce efficiently (he may instead prefer to maximise subsidy payments). Nor is there a direct test of whether the total economic value of his service exceeds its cost. Finally, the methods used to finance the subsidy will invariably give rise to inefficiencies elsewhere in the economy.

A more satisfactory solution is for the monopolist to set different prices for different customers. If he can discriminate between them, he can charge each a price which reflects willingness to pay for the good. Thus consumers with intense preferences will be charged a higher price than those with milder preferences.

Price discrimination is one way of overcoming the allocative inefficiency inherent in setting a price equal to average cost. If each consumer is charged according to his willingness to pay, the monopolist will be able to cover his total costs and his revenue will provide a direct measure of the demand for his services—which would not be so with a government subsidy. Price discrimination will also encourage the monopolist to expand output, thus reducing the social costs of monopoly.

The ability of a firm to charge consumers different prices for the same product depends on several factors. First, consumers must not be able to trade the product, otherwise those who bought it at a low price would sell it to those willing to pay more and pocket the difference. This cannot occur with

[1] A fuller discussion can be found in Kenneth D. George and John Shorey, *The Allocation of Resources*, George Allen & Unwin, London, 1978, pp. 120-31.

cable TV. Secondly, it must be possible to distinguish between consumers or groups of consumers with different preferences for cable TV.

Obtaining information about willingness to pay is costly and will limit a monopolist's scope for price discimination. Cable companies, however, will be particularly well-placed to discriminate because they will build up over time a substantial data base on consumer incomes, television viewing patterns, and the sensitivity of different groups of consumers to the price of different programmes. The present practice in the industry of selling cable TV in 'tiers' is a form of price discrimination. 'Pay-per-view', which Hunt recommended should be banned 'for the time being',[1] would also facilitate desirable price discrimination.

Whilst price discrimination of some sort is warranted on the ground of allocative efficiency, it nonetheless enhances monopoly profits. More of the consumer gain from cable TV is transferred to the cable operator as a result. Many will regard that as an undesirable cost of private monopoly to be set against the desirable allocative effects of price discrimination.

The problem restated

We can now restate the local monopoly problem more precisely. Laying cable requires a large fixed capital investment. The average cost of supplying an additional home with cable TV consequently declines as the number of subscribers increases. One cable system can therefore supply cable TV more cheaply than two or more, and there will be a natural tendency for monopoly supply in each area. The absence of competition in the market will enable the owner of a cable system to charge monopoly prices and, perhaps more significantly, control access to and the content of programmes on his system. This argument, however, applies only to the physical cable system; the supply of programmes and channels can still be organised competitively.

6. Do Cable Operators Have Market Power?

The existence of a local monopoly is, nevertheless, not sufficient to give cable operators market power. Several other

[1] Para. 50.

factors suggest rather that their market power will be much constrained.

(i) *Cable TV a price-sensitive luxury good*

First, cable TV differs from the services provided by traditional natural monopolies. Gas, electricity and water have no close substitutes. They are in an important sense essential services which endow their supplier with real and substantial market power. Cable TV, in contrast, is a *luxury* good and consumers will be sensitive to its price. Consequently, the scope for cable operators to raise prices significantly will be limited by the prospect of a rapid loss of customers and revenue.

(ii) *Potential competition in small catchment areas*

Secondly, the fixed investment required to lay a cable is much less than, say, to construct an electricty generator. The revenue needed to recoup the investment in a cable system is therefore also much lower. It is this factor which gives cable its local character; the optimal geographical catchment area is much smaller than in electricity generation and can be expected to fall as new technology becomes available and the market is able to support more cable systems.

The rapid pace at which technology is developing will also increasingly expose cable systems to potential if not actual competition. As their systems become obsolete, they will no longer be protected from entry.

Moreover, when the market for cable services expands and exceeds the capacity of the existing system, there is no reason why a second or new cable could not be laid by another company. Even though the two companies may not be able to co-exist profitably, competition between them, or even the threat of it, will inhibit the exercise of any market power that there may be. Competitive pressures will also come from contiguous cable systems which will have an incentive to undercut each other's prices in fringe areas to attract each other's customers.

(iii) *Many competitive non-TV services*

Thirdly, as the cable industry grows it will begin to offer a large variety of services, thus further distinguishing it from the traditional natural monopolies. The non-television services becoming available through interactive cable will have to compete with other types of communication and information services in markets which can be expected to be highly competitive.

For these three reasons, cable TV is not a typical natural monopoly requiring public utility status and/or extensive public regulation. It is unlikely that cable operators will enjoy significant market power under present or foreseeable market conditions.

SIX: Methods of Controlling Local Monopoly

In Section 5 we concluded that, although cable operators will probably have a local monopoly, they are unlikely to be able to exercise significant market power. It is admittedly a disputable assessment and one that Hunt (implicitly) rejects. The Report regards a totally free market in cable and TV as undesirable and recommends that cable operators be franchised and subjected to 'minimum constraints'.

In order not to pre-empt the issue, we now briefly examine several alternative approaches to the regulation of cable's local monopoly. Although our verdict on these alternatives is largely negative, the discussion will highlight their comparative advantages and disadvantages.

The principal alternative methods of controlling a local monopoly are public ownership, rate-of-return regulation, the separation of the ownership of cable systems from programming, and franchising.[1] The first will not be discussed in detail for several reasons. The principal one is that cable is not a classic natural monopoly and there is no reason why the public purse should finance what is a leisure good and perhaps a luxury. Further, public ownership does not by itself control market power; it must be combined with economic and financial controls to contain prices and ensure internal efficiency and good management.[2] An analysis of the difficulties of designing and enforcing these controls would take us well beyond the scope of this *Hobart Paper*.

[1] For two excellent discussions of the alternative methods of regulating local monopoly, Eli M. Noam, 'Towards An Integrated Communications Market: Overcoming the Local Monopoly of Cable Television', *Federal Communications Law Journal*, Vol. 34, 1982, pp. 209-57; and Richard A. Posner, 'Natural Monopoly and Its Regulation', *Stanford Law Review*, Vol. 21, February 1969, pp. 548-643.

[2] Ray Rees, 'The Pricing Policies of the Nationalised Industries', *Three Banks Review*, No. 122, June 1979, pp. 3-31; David Heald, 'The Economic and Financial Control of U.K. Nationalised Industries', *Economic Journal*, Vol. 90, June 1980, pp. 243-45.

1. Rate-of-Return Regulation

In the United States, private monopoly coupled with the regulation of the rate of return on capital by regulatory commissions has been the predominant method of dealing with a natural monopoly. Such public utility regulation has attempted to contain the twin evils of monopoly—excessively high prices and profits—by directly administering prices.

Public utility commissions invariably seek to control monopoly by establishing prices which permit a 'fair' rate of return on capital. Doing so requires an elaborate accounting system which will take into consideration all relevant factors and identify 'reasonable' costs.[1] It is not an easy task, for the monopoly firm is unlikely to be co-operative. Since the fair return is expressed as a percentage of capital costs, the firm has an incentive to inflate allowable costs in order to increase profits.

Drawback of investment distortion

The main drawback of this approach is that it runs a high risk of distorting the investment decisions of utilities. Economists have devoted considerable attention to the potential inefficiency of rate-of-return regulation, which has come to be known as the Averch-Johnson (A-J) effect.[2] If a public utility commission sets a rate of return on capital which is higher than its cost, it will encourage investment in capital-intensive production techniques. The public utility will not then employ the least-cost method of production but rather one with too much capital relative to labour and other inputs.

Rate-of-return regulation is fraught with such difficulties principally because it deals not with the cause of market power but its consequences. The utility retains its ability to manipulate other variables (such as product quality, service, introduction of new technology) to circumvent the rate-setting constraint. This approach also entails substantial adminis-

[1] The actual process of rate regulation is much more complex than this. For fuller discussions, Michael A. Crew and Paul R. Kleindorfer, *Public Utility Economics*, Macmillan, London, 1979, Chap. 8; Stephen Breyer, *Regulation and its Reform*, Harvard University Press, Cambridge, Mass., 1982, Chaps. 2 & 3.

[2] H. Averch and L. Johnson, 'Behaviour of the Firm Under Regulatory Constraint', *American Economic Review*, Vol. 52, December 1962, pp. 1,053-69. Also Crew & Kleindorfer, *op. cit.*, pp. 132-39; Rowley, *op. cit.*, Chap. 11.

trative complexities which are commonly underestimated by its proponents. The regulatory commission will, for example, have to pay considerable attention to deciding what are 'reasonable' costs and how to allow for inflation, risk, new technology, and changes in market conditions. In short, it will be continuously 'second-guessing' the monopolist's production and investment decisions.

Rate-of-return regulation is beneficial if it reduces prices and expands output. It would thus be unwarranted to conclude that it is *necessarily* undesirable. The increase in consumers' surplus resulting from higher output and lower prices must, however, be balanced against both the distortions described above *and* the administrative costs associated with 'price controls'. To these must be added the real danger that, over time, the regulatory agency will be 'captured' by the monopoly and will therefore serve the interests of the monopoly rather than those of the consumer.

As a means of regulating cable, this approach does not appear particularly attractive. There is simply insufficient information available about costs and revenue, and civil servants lack experience with this form of highly detailed regulation.

2. Mandatory Separation

A number of submissions to the Hunt Inquiry argued for a mandatory separation of ownership of the various functions of cable[1]—that is, that the cable *provider* (the company which installs and owns the physical plant) should be independent of the *operator* (the organisation which puts together and sells cable services to subscribers), and that the cable *operator* should be independent of the *programmer* (the assembler of programmes into channels or segments of channels for sale to operators).

The Hunt Committee, however, saw no need for separation. Its Report states that separation of operator and provider would seriously inhibit the financing of cable investment. Cable operators, adjudged by Hunt to have the main commercial motivation to install cable systems,[2] would find it

[1] For example, British Telecom, IBA, Centre for Policy Studies and National Consumer Council.

[2] Para. 22.

difficult to raise the required capital if they did not own and control the physical assets of the cable system and lacked direct control over the marketing of cable services.[1] That vertical integration between cable operator and programmer might lead to a restriction of programme choice was rejected by Hunt as being neither a serious nor a likely problem.[2] Hunt thus endorsed the conclusion of the ITAP Report that enforced separation 'would inhibit the development of the industry in its crucial formative stages'.[3]

Many other submissions to the Hunt Inquiry disagreed. They viewed the separation of functions as essential for controlling the local monopoly of cable operators, especially in facilitating disfranchisement by the regulatory authority. Ownership and programming are currently separated in commercial broadcast television. The IBA—a self-financing public body—builds, owns and operates all transmitters which it rents to the regional ITV programme companies. The rationale for separation was most clearly stated in the National Consumer Council's submission to Hunt:

> 'We do not see how there could be effective protection against the very real dangers of a monopoly of information supply if the functions of cable operation and programme provider are combined. Nor do we see how any regulatory authority could apply any ultimate sanction (for example, non-renewal of a licence) against a programme provider who also owned or operated the cable system in any area. In the last resort a combined cable operator and programme provider could not be closed down without the service being seriously damaged.'[4]

Common carrier status

Many submissions argued that the owners of cable systems should have common carrier status.[5] The idea behind common

[1] One submission argued cogently that separation would make financing difficult since cable operators would incur heavy deficits in the early years and investors and bankers would not provide finance without real assets as collateral: *Charterhouse Japhet plc Submission to the Hunt Inquiry,* May 1982, p. 8.

[2] Para. 23.

[3] *Cable Systems, op. cit.,* para. 6.22.

[4] 'Submission by the National Consumer Council to the Inquiry into Cable Expansion and Broadcast Policy', May 1982, para. 7.2.

[5] The Labour Party is in favour of a national common carrier operated by British Telecom.

carrier status is that the functions of cable operating and programming should be separated, and channels—or parts of them—should be made available on a first-come, first-served basis to anyone able and willing to pay the charge for access. The result, it is argued, would be competition and diversity in programming. If the operator controlled programming, it would be infected with the inefficiencies inherent in monopoly. Common carrier status would contain the monopoly problem in one allegedly manageable area, that of access, and ensure that the supply of programmes was competitive.

This proposal has an intuitive appeal and has been advocated by some economists.[1] It recognises the very important point that it is only the cable infrastructure that gives rise to cable's local monopoly.

Costs and benefits

It is difficult to assess the efficiency of this approach. It is not clear, for example, to what extent cable owners will also be programmers in the absence of enforced separation. Nevertheless, the claim that common carrier status would encourage programme diversity and expand consumer choice can be examined. Let us consider a system in which the operator was required to lease all his channels on a first-come, first-served basis at a uniform rental set by himself. He would set the rental at the level where the number of channels leased maximised his profits. Such a uniform rental would necessarily exclude some channels which could earn sufficient revenue to defray access and production costs, but insufficient to cover the rental. In contrast, the operator who was free to decide how his channels were used and the fees to be charged for them would continue to provide more channels so long as they added something to his profits, even if that amount was less than the uniform rental. Thus common carrier status would run the risk of excluding those very minority programmes (the marginally profitable ones) which cable is expected to provide.

Like most policies, common carrier status has costs and benefits. For those programmes which could earn sufficient revenue, the price to subscribers would be lower, the reason

[1] Elizabeth E. Bailey, 'Contestability and the Design of Regulatory and Antitrust Policy', *American Economic Review*, Vol. 71, May 1981, pp. 178-83.

being that competition between the programme companies would drive the price down to marginal cost. Thus common carrier status would not lead to an unambiguous reduction in consumer welfare. Neither, however, would it promote maximum choice for the consumer.

There are two other major costs associated with the common carrier proposal. First, if a local monopolist possessed significant market power, rate regulation would also be required. The problem of monopoly would merely be shifted from prices for subscribers to channel charges. It is not obvious that the calculations necessary to determine an efficient schedule of access charges would be any easier to make than those for viewers' subscription rates. Indeed the two are related.

Secondly, where separation did make a difference to the organisation of each local monopoly, it might increase costs. Much vertical integration can be explained as an effort to minimise the costs of dealing in the market. A vertically-integrated firm substitutes managerial decision-making for contractual agreements as a way of organising the different stages of production.[1] This may occur for two reasons: to enhance a firm's market power through control over key stages of production; or to economise on the costs of negotiating, forming, monitoring and enforcing contracts. If the second is the principal reason why operating and programming are in some cases controlled by the same firm, a policy of enforced separation would raise costs and subscriber charges. It would be a further factor leading to fewer programmes and channels for the viewer.

Conclusion on separation

The enforced separation of cable operators and programmers does not seem an attractive solution to the monopoly problem. It would, on balance, harm consumers by reducing the ability of operators to charge discriminatory prices and by raising the costs of the industry. The consequence would be less programme variety and a risk that the growth of the industry would be inhibited. If separation was to avoid diminishing programme choice, it would have to be coupled with controls on rentals for access which permitted discriminatory charges.

[1] Ronald H. Coase, 'The Nature of the Firm', *Economica* (new series), Vol. 4, 1937, pp. 386-405.

From here it would be but a short step to a full-blown regulatory system whose costs and inefficiencies must also be weighed in assessing the desirability of this approach.

The one appealing feature of the proposal is that it would reduce the price of the cable TV programmes shown. Whether that gain would be larger than the loss to consumers from reduced choice would depend on the degree of market power the operator possessed.

3. Franchising

Another approach to the control of local monopoly would be to award the franchise to the company which best satisfied specific criteria imposed by the franchising authority. This is the approach favoured by the Hunt Committee. Companies wishing to cable an area would have to submit bids describing the services they were willing to supply and other features of the cable system they planned to build. The franchise would be awarded to the cable operator which undertook to provide the 'best service for the area concerned'.[1]

The effects of franchising on market power and performance would depend on the conditions imposed on the cable operator and, more importantly, on how they were enforced. There is no reason to expect that franchising would enhance consumer welfare. The experience in the USA, where cities are primarily responsible for franchising, is that onerous conditions are often imposed and the whole process is open to corruption. Hunt proposes a new central cable authority to overcome some of these difficulties. In the absence of more detailed information about the proposed franchising agreement than are contained in the Hunt Report, however, it is difficult to assess the probable effects of this form of regulation. We now discuss franchising in general terms, leaving Hunt's more specific recommendations until Section 7.

Methods of franchising

Franchising can take many forms. The form recommended by Hunt would require cable operators to make competitive bids on the non-price aspects of the systems they proposed—the area to be cabled; the speed of construction; ownership,

[1] Para. 85.

number, range and diversity of channels; the amount of community programmes and local access to be offered; and intentions about interactive services.[1]

It would be by chance only if a franchising authority imposed conditions which maximised consumer welfare. For, like any regulatory body, it would lack appropriate incentives.

The Chadwick auction

There is, however, one method of franchising which deals directly with the consequences of monopoly. It is the so-called Chadwick auction of franchises.[2] Under this scheme, the exclusive right to supply a market is awarded to the company which undertakes to sell its goods and services at the lowest price.

The Chadwick auction relies on competition for the exclusive right to become the sole supplier to eliminate monopoly profits *and* prices. The applicants are told that the franchise will be awarded to the company which bids the lowest per-unit price to supply a service. Provided the auction attracts sufficient interest and collusion is avoided, competitive bidding will ensure that the price offered by the successful applicant is the lowest.

The Chadwick auction has been presented by some as a complete solution to the monopoly problem and a substitute for detailed regulation.[3] And, on the face of it, it does seem simple and attractive, overcoming the monopoly problem by competition in a way which benefits the consumer. But its simplicity and effectiveness are, unfortunately, more apparent than real.[4]

[1] Para. 86.

[2] Edwin Chadwick, 'Results of Different Principles of Legislation & Administration in Europe; of Competition for the Field, as compared with the Competition within the Field of Service', *Journal of Royal Statistical Society,* Vol. 22, 1859, pp. 22 ff.

[3] Harold Demsetz, 'Why Regulate Utilities?', *Journal of Law & Economics,* Vol. 11, April 1968, pp. 55-65. Also George J. Stigler, *The Organisation of Industry,* Irwin, Homewood, Ill., 1968, pp. 18-19; Richard A. Posner, 'The Appropriate Scope of Regulation in the Cable Television Industry', *Bell Journal of Economics & Management Science,* Vol. 3, Spring 1972, pp. 78-129.

[4] For a critical evaluation, Oliver E. Williamson, 'Franchise Bidding for Natural Monopolies in General and with Respect to CATV', *Bell Journal of Economics,* Vol. 7, Spring 1976, pp. 73-104; Richard Schmalensee, *The Control of Natural Monopolies,* D. C. Heath, Lexington, Mass., 1979, pp. 67-73; H. M. Trebing, 'The Chicago School *versus* Public Utility Regulation', *Journal of Economic Issues,* Vol. 10, March 1976, pp. 97-126.

Franchise specification and management

All methods of franchising entail difficulties, particularly those inherent in the initial specification of the franchise since firms can obviously bid only if the conditions are clearly defined. In the competitive bid scheme envisaged by Hunt, minimum conditions would be determined by the franchising authority and each applicant would seek to put together an attractive package of services to maximise the likelihood of winning the franchise. The franchising authority would then evaluate the bids and choose the one it considered the 'best'. In theory, the franchising authority could seize on a set of conditions which improved consumer welfare. Since, however, it would have no data on the impact of different terms and very little incentive to promote the interests of consumers (if only because their identity would not be known before the franchise was awarded), the franchise agreement could well exacerbate the problem of local monopoly.

The Chadwick auction seems to overcome this problem. The discretion of the franchising authority is apparently limited to awarding each franchise on the basis of one easily-measured criterion. In reality, however, this cannot be so since to award the franchise solely on the basis of the lowest price to subscribers requires that all other terms be identical. If each applicant can offer a different package of services, price alone will not provide an adequate criterion for making the award. An offer of six channels of high-quality programmes at £20 a month cannot readily be compared with one of 20 channels with a lower subscription rate. The lowest price ceases, therefore, to be an adequate criterion for awarding the franchise if the other terms of the offer may also vary. The Chadwick auction thus gives the franchising authority the same discretion over the non-price terms as other franchising methods with, again, no guarantee that they will maximise consumer welfare.

The comparability defect has been recognised by Professor Richard Posner who has proposed that subscriber preferences be determined by applicants as part of the bidding process. He has suggested

> '. . . an "open season" in which all franchise applicants were free to solicit the area's residents for a set period of time. This would not be a poll; the applicants would seek to obtain actual

commitments from potential subscribers. At the end of the solicitation period, the commitments received by the various applicants would be compared and the franchise awarded to the applicant whose guaranteed receipts, on the basis of subscriber commitments, were largest. In this fashion the vote of each subscriber would be weighted by his willingness to pay, and the winning applicant would be the one who, in free competition with the other applicants, was preferred by subscribers in the aggregate. To keep the solicitation process honest, each applicant would be required to contract in advance that, in the event he won, he would provide the level of service, and at the rate represented, in his solicitation drive.'[1]

For this procedure to work, however, there must be confidence that potential customers can assess cable TV and make their purchase plans in the abstract. Since consumers in Britain have no experience of cable TV, they are highly unlikely to be able or willing to commit themselves to purchases two or three years ahead—particularly at a time of recession.

Overbidding

Franchise applicants also have an incentive to make excessively attractive bids in order to have first crack at the market. There are significant advantages in being the first to cable an area. Once capital is in place and experience has been built up, a decision to disfranchise an existing operator will be costly. Unless the franchising authority has powerful means of enforcing the terms of the franchise, applicants will make lavish bids which they do not intend to honour.[2]

Overbidding is thus a serious risk with competitive-bid franchising. Once a franchise has been won and construction of the system begun, the successful applicant will start lobbying the franchising authority for extended completion dates, price increases, and the modification of technical standards, coverage requirements and service quality. It will not be short of arguments for modifying the original agreement, such as cost increases, honest mistakes in estimation, and low take-up

[1] Richard A. Posner, 'The Appropriate Scope of Regulation in the Cable Television Industry', *Bell Journal of Economics & Management Science,* Vol. 3, Spring 1972, pp. 98-129, extract from p. 115.

[2] As one commentator has observed, 'cable companies have been promising the moon; cities have been demanding the earth'. (Robert MacNeil, 'Cable Klondike', in Brian Wenham (ed.), *The Third Age of Broadcasting,* Faber & Faber, London, 1982, p. 107.)

rates. The experience of the City of Oakland, California, demonstrates the scope for manoeuvre.[1]

Even when overbidding is negligible, negotiation will still be required. A franchise is essentially a long-term contract between the cable operator and the franchising authority which requires both parties to commit themselves to long-term plans. Over time, as new circumstances unfold and significant changes occur in costs, demand and technology, some of the original terms will cease to be profitable and performance will fall short of promise. Adaptation to change will necessitate re-negotiation, arbitration and agreed rate formulae. If, however, re-negotiation is implicit in franchising, it calls into question its superiority over detailed regulation in securing the best service for the consumer. Hidden in the arrangement is an extensive system of 'contract management' by the franchising authority.[2]

4. The Balance Sheet of Regulation

All forms of government intervention have benefits and costs. The effects of the most frequently-used methods of controlling local monopoly have now been discussed. The balance sheet is far from comprehensive because the data are lacking to permit firm quantitative estimates. What is evident is that all the methods entail significant costs both to the industry and to the taxpayer. Many are 'hidden' costs in the form of distortions and reduced productivity. The costs can, however, be justified if they are exceeded by the benefits to consumers which regulation may produce.

We suggested in Section 5 that the market power of cable operators was unlikely to be significant and hence that the benefits of regulation would be small. If, however, some form of regulation is thought necessary, it must be designed to be cost-effective. Cost-effectiveness can be maximised by relying on market-like methods of control such as pricing schemes and pro-competition policies. It is the last which, we shall argue later, should form the basis of government policy on cable TV.

[1] Williamson, *op. cit.*, pp. 92-103.

[2] W. M. Crain and R. B. Ekelund, 'Chadwick and Demsetz on Competition and Regulation', *Journal of Law & Economics*, Vol. 19, April 1976, pp. 149-62; Victor P. Goldberg, 'Regulation and Administered Contracts', *Bell Journal of Economics*, Vol. 7, Autumn 1976, pp. 426-48.

SEVEN: The Hunt Report

This Section examines the approach and recommendations of the Hunt Report.

1. Hunt's Basic Approach

The basic approach of Hunt is that 'cable television is about widening the viewer's choice'[1] and that it 'should be seen as supplementary, and not as an alternative or rival, to public service broadcasting'.[2] In Hunt's view, cable TV cannot be completely free of regulation because it will not be available to the whole country, because there is as yet no general consensus about what should be shown on cable, and because the cable operator will have an 'effective local monopoly'. Nor, says Hunt, can there be sole reliance on self-regulation in the formative years because views within the industry are likely to differ substantially. Cable operators must therefore be required to observe 'certain liberal ground-rules'.[3] Hunt's basic concept is that they

> 'should be accountable to observe a few general guidelines and provide the service offered when given their franchises rather than one of a central body regulating in detail how they go about their business.'[4]

The emphasis is on 'oversight' rather than 'detailed regulation'. As Hunt points out, the information necessary to frame a rational scheme of detailed regulation is simply unavailable: 'cable television is a leap in the dark'.[5]

The Report reflects the policy of the Thatcher Government that the expansion of cable television should begin as soon as possible and with 'minimum constraints'. As a *Sunday Times* leader concluded, cable TV after Hunt

> '. . . is to be financed, as is the way of Mrs Thatcher's Britain, with private money. . . . That is the premise of the Hunt Report,

[1] Para. 9. [2] Para. 8. [3] Para. 12.

[4] *Ibid.* [5] Para. 13.

and the underlying reason why the ground-rules he proposes are liberal to the point of invisibility.'[1]

Whether these ground-rules will in practice be 'liberal to the point of invisibility' is debatable. Both Hunt and the Government have proposed measures to regulate the industry which, if implemented, could severely restrict free competition.

The Hunt Report can be neatly divided into two parts corresponding to its remit: (a) the protection of broadcast TV; (b) the regulation of the cable industry.

2. Protecting Public Service Broadcasting

The Hunt Committee was required by its terms of reference 'to recommend measures to protect public service broadcasting' in a way consistent with the wider public interest. Its Report is to be applauded for placing very few restrictions on cable to protect the BBC and ITV. Indeed, its recommendations that cable be permitted to accept advertising and to relay out-of-area ITV programmes will increase competitive pressures on ITV programme companies, thus weakening their regional monopolies.

Hunt believes that some 'limited safeguards against damage to public service broadcasting' are necessary. It recommends three measures:

1. *A ban on pay-per-view:* cable operators will not be permitted to levy charges for individual programmes.
2. *An 'anti-siphoning' rule:* cable operators will be denied exclusive rights to national sporting events.
3. *A 'must-carry' rule:* all BBC and regional ITV broadcasts must be relayed by cable TV.

For and against siphoning

The ban on pay-per-view and the 'anti-siphoning' rule are both designed to prevent a local cable operator from obtaining the exclusive rights to a national sporting event by outbidding the BBC and ITV. Without offering reasons or examining the full implications of these restrictions, Hunt simply assumes that such events should be made available to all viewers.

[1] *Sunday Times,* 17 September 1982.

Some national sporting events, for many of which BBC and ITV pay relatively little, are now offered 'free' to the public over broadcast TV. If cable operators were allowed to bid for them, all viewers would still receive them 'free' provided BBC and ITV were willing to pay at least as much as cable operators. Hunt gives no convincing reason why BBC and ITV should not pay the full market value for the right to these events; on the contrary, it expresses 'some sympathy with those who suggested to us that television at present secures sporting coverage on the cheap'.[1] No-one has yet argued that Universal Pictures should be prohibited from selling its film *ET* to pay-per-view audiences in cinemas so that BBC or ITV can get the film at a bargain price. And the Hunt Report itself rejects the cinema industry's demand for an exclusive right to new films during their first 12 months,[2] stating unequivocally 'that market forces should be allowed to dictate the showing of films on cable'.[3] Are national sporting events different?

The strongest argument against siphoning is that, for some events, pay-TV would have the effect of transferring wealth—a purely distributional issue. In the FA Cup Final, for example, footballers who are already well paid would receive even higher sums, as would shareholders of football clubs. But the result would be neither more nor better-quality football. All consumers of television football programmes would be worse off, some because they would now have to pay and others because they could no longer view or because they found viewing too expensive. By no means in all circumstances, however, would subjecting sporting events to pay-per-view merely transfer wealth from viewers to sports clubs and cable companies. The stretched finances of many sports clubs today necessarily constrain the quality of their output, which they could be expected to improve with additional revenue from selling telecasting rights to cable in a more competitive market. Here it is odd that Hunt did not recognise the adverse consequences for sport of a broadcasting monopoly.

Cost-benefit test of effects of siphoning

The economic effects of siphoning can be appraised by a cost-benefit test. The cost is the loss in consumers' surplus from

[1] Para. 70. [2] Para. 80. [3] *Ibid.*

having to pay for or not to be able to view a sporting event which would otherwise be broadcast free. The benefit (if any) is the gain in consumer welfare from more and better programmes as a consequence of the higher returns to sports clubs and event organisers. Thus combatting the 'harmful' consequences of siphoning requires a selective ban in those few instances where

(a) BBC or ITV wish to telecast the event, *and*

(b) the sole or main result of siphoning would be to redistribute substantial welfare away from consumers.

A provision such as in Section 30 of the Broadcasting Act of 1981 would be appropriate, permitting the Home Secretary to designate a sporting or other event as one of national importance to be reserved for broadcast TV. In addition, the Home Secretary should be required to state his reasons, should be authorised to act only when BBC or ITV genuinely wish to broadcast the event, and should be empowered to fix the fee if the 'owners' of the event complain of unfair deprivation of earnings.

Pay-per-view ban

Satisfying the cost-benefit test requires only a *selective* anti-siphoning rule. It certainly does not require a ban on pay-per-view as well. Hunt justifies this additional measure by arguing that cable operators would be more likely to outbid ITV or BBC for the right to televise national sporting events if they could levy a separate charge.[1] Is this true? And what of the possible misallocative effects of the ban on other programmes?

In Section 4 we argued the merits of discriminatory pricing to make revenue a more reliable measure of consumer preferences and to avoid monopoly restrictions on output. Yet, in order to give broadcast TV a monopoly of national sporting events, the cable industry is to be prohibited from using an attractive and efficient method of pricing TV programmes. Moreover, when pay-per-view becomes a feasible method of pricing, the ban will have adverse consequences for minority programmes. Let us suppose the Glyndebourne Opera Company were to offer an appealing production on a pay-per-view basis. Not only would the viewers of the programme

[1] Para. 50.

benefit, but the additional revenue generated from selling the TV rights would contribute to the company's financial stability. The consequence would probably be more opera productions in the future, from which all opera lovers would gain. *A ban on pay-per-view here would be unambiguously harmful.*

The principal objection to the ban is that it will not by itself prevent cable TV from purchasing exclusive rights to national events. The cable programme companies will still be able to outbid the BBC and ITV even if they are not allowed to charge separately. The ban is, in short, ineffective because it will not achieve its stated aim and undesirable because it may lead to avoidable inefficiencies in programming. There is also a slight danger that it will inhibit the growth of interactive technology. One of the reasons for installing interactive cable is that it can be used for metering programme consumption on a pay-per-view basis. If the latter is prohibited, the incentive to invest in an interactive system will obviously be smaller. The significance of this effect is, however, impossible to quantify.

The 'must-carry' rule

The proposal to impose a statutory obligation on cable operators to relay all BBC and regional ITV channels is obviously designed to shield them from competition. It is unlikely, however, to have any significant practical effect since cable operators will probably find it in their own interests to offer these channels to subscribers as part of the 'basic tier'. Only existing cable operators who have out-of-date systems with a maximum of four channels will be adversely affected. For these systems Hunt recommends that the 'must-carry' rule be waived for up to five years *provided* their operators offer viewers an alternative means of receiving broadcast TV free of charge.[1]

The safeguards assessed

The safeguards for public service broadcasting which Hunt recommends are exceedingly mild if not irrelevant. Reading between the lines of its Report, the message is clear: broadcast TV should not be protected from competition. The only safeguard which should be seriously reconsidered is the 'pay-per-

[1] Para. 59.

view' ban—a reconsideration which, at the time of writing, the Home Secretary has announced is taking place.[1]

3. The Supervisory Framework

The second and more important part of the Hunt Report is devoted to how the cable industry should be regulated. It rejects public ownership, rate regulation and mandatory separation of owner and operator as viable policies. Public ownership is dismissed because it is inconsistent with the Government's view that 'cabling should not make significant demands on public expenditure'.[2] And detailed regulation is rejected because it would be 'impossible without a large bureaucracy which would stultify . . . initiative and diversity'.[3] Hunt sees 'no need to control charges'.[4] It recommends instead that cable operators be franchised by competitive bidding to secure the 'best service for the area concerned'. Franchising would be administered by a new central cable authority with two roles: awarding the franchises and 'general oversight' to ensure that operators honour their terms and observe other ground-rules. As Hunt states:

> 'General oversight . . . will provide the opportunity to respond flexibly as the industry develops in ways which are impossible to forecast now.'[5]

4. Hunt's Demand for Regulation

The Hunt Report can be searched in vain for a convincing reason why the industry should be regulated. An early paragraph gives the impression that regulation is required because film is a powerful medium and because less than the whole country will have access to cable TV.[6] Later on it is the cable operator's local monopoly which is said to necessitate regulation.[7]

The word 'monopoly' is mentioned no less than 10 times in the Report. Yet nowhere is it stated why the local monopoly of cable operators should be controlled and what undesirable

[1] Peter Riddell and Jason Crisp, 'New Authority will Supervise Expansion of Cable Television', *Financial Times*, 3 December 1982, pp. 1 & 8.

[2] Para. 22.
[3] Para. 7.
[4] Para. 92.
[5] Para. 13.
[6] Para. 11.
[7] Para. 18.

consequences it will have, let alone how franchising is going to deal with them. What fleeting discussion there is merely suggests that local monopoly is necessary to secure the financial viability of the cable companies—and thus is presumably a good thing. Moreover, Hunt's specific recommendation *against* controlling charges to cable customers implies that it does not consider monopoly pricing will be a serious problem.

The Report contains no analysis of the case for regulation. Nor does it describe what evils the cable authority is supposed to protect the public from. It is difficult, therefore, to accept its view that franchising is necessary.

Hunt makes the elementary and fundamental mistake of assuming that local monopolies in cable TV render a free market undesirable. As we have sought to emphasise, cable operators can exploit the consumer only if they enjoy a monopoly in both the geographical *and* the product markets. Hunt fails completely to take this into account. Why will the market not secure the best service for the consumer? How will the franchising authority do a better job? Why cannot existing UK and EEC competition law deal with whatever manifestation of market power may emerge? All these questions are ignored in the Report.

Ironically, the only purpose for which the Report makes out a convincing case for centralised regulation is to deal not with any deficiency in the market but with the potential excesses of government—namely, the risk that local authorities will use their power to grant wayleaves (that is, to give permission to dig up roads) as an opportunity to impose onerous and unnecessary conditions on the cable industry.[1] Hunt sensibly recommends that the new cable authority should have either the power to grant wayleaves directly or else the reserve power to override the decisions of local authorities.[2]

5. Specific Recommendations

Hunt makes a number of specific recommendations about restrictions on the ownership of cable systems and the conditions to be included in franchises.

[1] An example is mentioned by Hunt. In the USA it is the practice of municipalities to charge a franchise fee of between 3 and 5 per cent of cable operators' revenue. These fees have the undesirable effect of exacerbating monopoly prices.

[2] Para. 92.

Restrictions on ownership[1]

The Report recommends very few limitations on the ownership of cable-operating companies. It proposes that central and local governments, political parties and organisations, and religious bodies be banned from owning operating companies; and that rival press, independent television and local radio contractors, and foreign companies should not have a controlling interest in them.

The Report justifies the exclusion of government bodies, political parties and religious institutions on the ground that cable-TV operating companies 'should be free from any kind of political or ideological bias'. This ban seems sensible because, as non-profit organisations which do not have to respond to market incentives and consumer preferences, they will allow non-commercial reasons to influence programme policy and determine access to the system. Even the religious fanatic, however, is not wholly immune from market forces. People will subscribe to cable TV only if they regard it as a good buy, which it would not be if it screened 24 hours a day of 'bible bashing'. It is true that political and religious organisations could subsidise their systems out of revenue generated by their other activities. But they do not have unlimited finances, and cable is not a cheap way to 'buy' votes or converts.

Foreign companies should be banned from owning a controlling interest, according to Hunt, because that is 'unacceptable'.[2] Unacceptable to whom, and why, in terms of the costs and benefits, is not explained. Moreover, the recommendation is ambiguous. Only *individual* foreign companies are to be forbidden a controlling interest, thus not ruling out the possibility of complete control by two or more.[3] The restriction will, nevertheless, deter foreign capital from investing in cable expansion. Since the investment required will be both large and risky, and since US capital has far more experience of this form of investment, the harmful consequences may be substantial.

[1] For a comprehensive discussion of cable ownership, Kenneth Gordon, Jonathan D. Levy and Robert S. Preece, *FCC Policy on Cable Ownership—A Staff Report*, Federal Communications Commission, Washington DC, November 1981.

[2] Para. 26.

[3] Compare para. 26 with the more explicit wording in para. 9 of the 'Summary of Conclusions and Recommendations'.

Reducing danger of cross-media monopoly

Restricting ownership by the press and radio and television companies makes more sense. Once again, however, the Report justifies by assertion: '. . . control by another media interest could lead to undesirable monopoly power'.[1] This result is certainly possible since press, radio and broadcast TV will all be in competition with cable TV. If one firm controlled all four media in the same geographical area, it would enjoy some market power provided barriers to entry were significant. And if a regional ITV company owned the local cable-TV operating company, it would have an incentive to suppress those programmes which attracted away a significant proportion of its viewers. In deciding on the profitability of showing a particular programme on cable, the TV company would take into account the consequent loss of advertising revenue from its broadcast TV operations. Where such audience diversion was significant, the net profitability of pay-cable would be lower to such a company than to an operating company with no interests in broadcast TV.

Again, however, merely identifying this potential restrictive practice says nothing about the magnitude of the harm from cross-media ownership. Studies in the USA suggest that the audience diverted from broadcast to cable TV is not large.[2] Even if the risk is substantial, Hunt should have explained why existing competition legislation could not deal with it effectively. The ban is nevertheless probably sensible on balance since its only effect will be to reduce the likelihood that cross-media ownership will restrict competition.

Franchise conditions[3]

Hunt gives few details of the conditions to be imposed on cable operators. And its discussion is confusingly ambiguous because many of the conditions are *suggested* rather than *recommended*. What weight Hunt intends its *suggestions* to carry

[1] *Ibid.*

[2] Rolla E. Park, *Audience Diversion Due to Cable Television: A Statistical Analysis of New Data*, Rand Corporation, Santa Monica, R–2403–FCC, April 1979, and references cited.

[3] We have already discussed the difficulties with competitive bidding for franchises and shall not repeat them here.

in the Government's final decisions about cable policy is left to the reader's imagination. It cannot be much.

The Report is also ambiguous about whether franchises should be exclusive. It implies that they should be, thereby introducing an unnecessary legal support for the cable operator's local monopoly. As Professor Kahn has stated: 'No barrier to entry is more absolute than one imposed or enforced by the sovereign power of the state'.[1] If local monopoly is a reality, statutory exclusivity is unnecessary; if it is not, exclusivity will create market power considerably in excess of that made possible by economies of scale. This restrictive and anti-competitive aspect of exclusive franchises seems to have eluded both Hunt and its critics. Together with the Report's failure to make out a sufficient case for franchises, this oversight must call into question all its detailed proposals on franchising. We shall, nevertheless, examine each separately.

Limited areas

Hunt *suggests* that franchises should limit each cable operator's area to less than half-a-million homes so as to be small enough to identify with the local community. It is difficult to see why this objective, which is more appropriate for community cable, should be foisted onto commercial companies.

The optimum commercial scale of a cable system is determined not by its capacity to identify with a geographical area but by the economics of efficient operation. An arbitrary limitation on the area a cable operator can serve will damage its financial viability or else harm consumers by pushing subscriptions higher than strictly necessary. The Hunt Report recognises that there is a trade-off between the size of a cable system and its operator's financial viability—which probably explains why it declines to *recommend* an arbitrary limit. What it does not seem to appreciate is that restricting franchise areas to promote the goal of local identity will be bought at the cost of higher prices for cable TV and reduced competitiveness with other forms of TV and telecommunications.

'Cherry-picking'

It is also *suggested* in the Report that the new cable authority

[1] Alfred E. Kahn, *Economics of Regulation,* Vol. I, Wiley & Sons, New York, 1971, p. 116

should encourage cross-subsidisation of cable TV in different areas. The discussion of this topic loses its hold on economic reality in a garbled analysis of how cable expansion can be accelerated:

> '. . . if cable is to spread successfully, many relatively small towns should be cabled even though they may not be able to support such a large number of channels. . . . The franchising body will need to take into account the comprehensiveness of the cabling within the area concerned, so that it is not restricted to streets where a majority of occupiers are likely to pay for cable. It could also speed up the widespread development of cable and avoid what is known as "cherry-picking" by seeking in some cases to combine less attractive localities with those carrying the best commercial prospects or even by asking large consortia which bid for a prosperous area to provide a separate cable system in a less promising area elsewhere.'[1]

This proposal is the most unattractive and restrictive put forward in the Report. It appears to be based on a belief that everyone is entitled to cable TV and that this can be realised by forcing the industry as a whole to subsidise cable expansion in unprofitable areas. If it is a serious suggestion, it would not, as Hunt naïvely claims, 'speed up the widespread development of cable'. Rather it would inhibit its expansion by imposing a cost on the industry equivalent to a tax, raising subscriptions in profitable areas and distorting the allocation of resources by encouraging cable TV in areas where its cost would substantially exceed the benefit to consumers. Why a luxury good should be made available to those who cannot pay its full cost ought to be explained before an industry is compelled to shoulder the burden of such an ill-thought-out policy. This omission reflects a weakness which infects the whole Report—its failure to evince an understanding of the market system in general and the cable industry in particular. As a result, it proceeds from recommendation to recommendation by the force of assertion rather than reasoned argument.

Duration of franchises

The Report *recommends* that the cable operator's franchise should initially be granted for 10 years, and thereafter for the same eight-year period as the franchises awarded to the ITV programme companies.

[1] Para. 86.

It has been argued that eight years are far too short to recoup the installation costs which are normally written off over 25 years. Indeed, if eight years are 'optimal' for ITV programme contractors, which are not required to invest in transmitters, it is scarcely arguable that they also constitute an appropriate period for cable operators. The Government has announced its intention to award franchises for a longer period: 12 years initially for operators and also 12 or 20 years for cable providers depending on the technology they employ.[1]

The principal objection to Hunt's recommendation is that, because franchises do not need to be exclusive, they therefore do not need to be renewable. The sole effect of exclusive franchises of a given duration is to protect the sunk capital of incumbent operators from competition from those willing to invest in newer and cheaper systems in the future.

Given exclusive franchises, however, one important factor must be borne in mind in determining their optimal duration. The duration of a franchise does not fix the period over which an operator must recoup his capital since, if he has performed in accordance with its conditions, the franchise may be renewed. It is the uncertainty surrounding renewal which will give rise to anxiety on the part of investors. If a particular operator foresees difficulties, he will have an incentive to write off his investment quickly and seek a higher rate of return to compensate for the risk of non-renewal. This will inevitably impose higher prices on the consumer. In addition, as the end of the franchise approaches, and with it the prospect that the owner-operator will lose it, he may be tempted to cease maintaining the system properly.

Solution to franchise non-renewal disincentives

Hunt does recommend one measure to deal with these disincentive effects. If an operator who also owns the cable loses his franchise, he should be required 'to sell or lease his infrastructure on a predetermined basis to another operator'.[2] This solution would be effective *provided* the assets were valued at their true market value. The original owner-operator would then be indifferent as between continuing or ceasing to operate

[1] Mr William Whitelaw, Home Secretary, in the House of Commons on 14 December 1982.

[2] Para. 87.

and would have no incentive to write off his investment at an accelerated rate. Moreover, the measure proposed by Hunt would substantially reduce barriers to entry. Compelling the original owner-operator to transfer his assets would eliminate the advantage his sunk costs would give him over applicants for his franchise who would otherwise have to install new plant.

A full assessment of the likely effects of the measure proposed by Hunt cannot be undertaken because its Report does not indicate how assets should be valued. Experience with regulation in the USA suggests that asset valuation is not straightforward.[1] A simple rule that a cable system should be valued at original cost minus depreciation can easily be manipulated by the franchise holder. But the main difficulty is the one discussed in Section 6 (p. 80). Third-party valuation of assets will have potentially distorting effects on the investment and pricing decisions of the owner-operator since such decisions will be made with an eye not only to costs and market conditions but also to the compensation formula—perhaps with the intention of making it very expensive for a rival bidder to take over the existing plant. Hunt's proposal would, nonetheless, encourage competition at the stage of franchise renewal and ensure continuity of service to subscribers.

6. General Oversight by Cable Authority

According to Hunt, once franchises have been awarded the cable authority should be responsible

> 'for judging whether cable operators were living up to their promises and for responding in a flexible way as the cable industry developed'.[2]

Its approval would be required to modify the conditions of the franchise; it would act as a forum to advise operators; and it would receive and adjudicate complaints from consumers.

These roles for the cable authority would appear to be relatively straightforward and to pose no real difficulties. Such a view, however, ignores two factors. First, the dangers of overbidding and the consequent need for 'contract management' emphasised in Section 6 (p. 88) will be exacerbated if the cable authority imposes unprofitable terms such as 'no cherry-

[1] Williamson, *op. cit.*, pp. 84-90.

[2] Para. 89.

picking'. Secondly, franchises will inevitably have to be re-negotiated since operators cannot be expected to be tied down to inflexible investment plans for such lengthy periods, given the considerable uncertainty about cable's market potential.

The real possibilities of overbid and subsequent re-negotiation call into question Hunt's claim that franchising can 'secure the best service for the area concerned'. If franchising is to have a concrete effect, it will necessarily entail detailed regulation. Otherwise it cannot guarantee that the operator initially awarded the franchise will be the best—especially when the penalities for breaching the agreement are rather crude. Hunt recommends the loss of franchise for 'gross breaches' and the imposition of a 'regulatory régime' for less serious infractions. A regulatory régime would require an operator to submit his programme schedules or would impose advance vetting of programmes or advertising. Hunt rejects performance bonds and other financial penalties because they are 'not well suited to failings which are a matter of qualitative judgement [such] as breaches of taste and decency'.[1]

7. The Cable Authority

The Report recommends a new authority to award franchises and oversee the cable industry. The Government has accepted this recommendation in principle, thus agreeing to the creation of a second quango in the communications sector—alongside an Office of Telecommunications (OFTEL) already planned to oversee the privatisation of British Telecom.[2]

'Capture' theory of the quango

It is one thing to theorise about the potential benefits of regulation, but quite another to ensure that a regulatory agency promotes the public interest in practice. The work of the new Nobel Laureate in Economic Science, Professor George Stigler, has warned emphatically against assuming that government will protect consumers.[3] He maintains that there

[1] Para. 101.

[2] *The Future of Telecommunications in Britain, op. cit.*, p. 2.

[3] George J. Stigler, 'The Theory of Economic Regulation', *Bell Journal of Economics and Management Science,* Vol. 2, Spring 1971, pp. 3-21; also his *The Pleasures and Pains of Modern Capitalism,* 13th Wincott Memorial Lecture, Occasional Paper 64, IEA, 1982.

is a 'market' in the formulation and implementation of government regulation which is overly responsive to producer interests. In general, regulation has a tendency to favour politically influential sectors of industry and to be designed to redistribute wealth in their favour, usually by restricting competition.[1]

Even when there is no such political motivation for new regulation, 'quangos' often develop a cosy relationship with those they control. The so-called 'life-cycle' theory of regulatory behaviour suggests the following pattern. In its early years, a quango faithfully carries out its statutory duties and vigorously enforces the law. As it grows older, its officials get to know the people in the industry it regulates and become sympathetic to their interests. The problems of the industry become the problems of the quango, and a much more lax and accommodating form of control evolves. Whether by design or by progressive development, regulators are gradually 'captured' by the industry they regulate. They end up promoting its interests rather than those of the consumer.

The casual observer of the UK economy will have no trouble identifying examples of regulation in the interests of the regulated and against the interests of the consumer. He will readily recall the long campaign by the Civil Aviation Authority (abetted by Ministers) to bar the Laker Skytrain and thereby maintain high prices on the Atlantic route in the interests of its large, inefficient, long-established client, British Airways. Or he will reflect upon the era of regulation of long-distance bus and coach routes which kept fares enormously higher than they have become since de-regulation in 1980.[2]

Even if the case for franchising cable is accepted, it is difficult to come to a firm conclusion whether the industry should be supervised by its own authority, as Hunt proposes, or by an authority responsible for communications in general, like the FCC.

Incentives to minimise risks of 'capture'

Theory and practice counsel caution and emphasise the desirability of building into the supervisory framework incen-

[1] William A. Jordan, 'Producer Protection, Prior Market Structure and the Effects of Government Regulation', *Journal of Law and Economics*, Vol. 15, April 1972, pp. 151-76; Richard A. Posner, 'Theories of Economic Regulation', *Bell Journal of Economics and Management Science*, Vol. 5, Autumn 1974, pp. 335-58.

[2] John Hibbs, *Transport without Politics . . . ?*, Hobart Paper 95, IEA, 1982.

tives to minimise the risks of 'capture' and the abuse of discretion by the quango. One method of minimising the risks is by establishing appropriate organisational arrangements within the regulatory agency. The common practice in the civil service, for example, of rotating personnel every two years reduces the likelihood that an official will come to identify with a particular industry. Giving a quango responsibility for several industries also helps. An authority specifically and solely for cable would run a bigger risk of 'capture' by the industry and of inclining over time to protect incumbent owners and operators against new competition and new technology. The Pilkington Committee on Broadcasting identified the same problem in describing the IBA's predecessor (the ITA) as a 'friend and partner' of the ITV programme companies.[1] A single agency for the communications industry as a whole would be more vigorous in promoting the general interests of the industry and in resisting attempts to impose restrictions on cable to promote broadcasting.

A broadly-based regulatory quango—a British Communications Office, say—would succumb less easily to 'capture' because it would have competing client industries. Were it to be given increasing responsibility over time for new industries, however, it might tend to favour old technology over new. Experience with the FCC suggests this danger. Hunt's recommendation that neither the Home Office nor the IBA should be given control over cable because of the risk of a conflict of interests indicates an awareness of the pitfalls. The IBA would inevitably adopt an over-protective attitude to public service broadcasting which, Hunt believes, 'could deter potential investors'.[2]

There is no simple answer to the problems posed by regulation once it is accepted as necessary. The discretion of the cable authority can be reduced to a minimum and a system of checks and balances established to help ensure its decisions are consistent with the Thatcher Government's policy of enhancing consumer welfare and improving efficiency.[3] Since,

[1] *Report of the Committee on Broadcasting*, Cmnd. 1753, HMSO, 1960, para. 572.

[2] Para. 95.

[3] In the USA President Reagan has ordered all executive regulatory agencies to prepare cost-benefit studies of proposed regulations. They are prohibited from undertaking regulatory action 'unless the potential benefits to society from the regulation outweigh the potential costs to society'. (Executive Order 12291 of 17 February 1981, *Federal Register*, Vol. 46, 19 February 1981.)

however, no convincing case for regulating cable has been made out, the risks and drawbacks of intervention must outweigh the undefined and probably non-existent benefits.

8. Conclusions

The Hunt Report is a document of considerable significance because it makes a sharp break with 60 years of government policy designed to restrict competition in broadcasting. Hunt's importance lies in its rejection of the argument that public service broadcasting should be shielded from competition from cable. On the other hand, its discussion and advocacy of regulation are unsatisfactory and weaken its otherwise sensible 'hands-off' stance.

The big disappointment of the Report is that it misses an opportunity to clarify the issues surrounding cable expansion. Its analysis is shallow and often incoherent. Its recommendations are all too frequently asserted rather than established by reasoned argument. Even the fundamental question of why cable's local monopoly should be regulated is not discussed, let alone answered. This part of the Report is a typical example of British compromise—and as such intellectually muddled. If three educated and experienced men, after spending six months of intensive investigation and £47,418 of taxpayers' money, were unable to establish a credible case for regulating cable, it would seem reasonable to conclude that none exists.

EIGHT: Principles and Proposals

In the light of the preceding analysis, we now outline the principles—and a supervisory framework to promote them—that in our view should form the basis of government policy on cable.

1. Basic Principles

Video publishers

The Hunt Report rejects the thesis that cable TV is 'just another branch of publishing' on the ground that less than the whole country will have access to it. This reason obscures the central feature of cable TV which makes it akin to publishing. Unlike traditional broadcasting, pay-cable is in essence a private relationship between a subscriber and a cable operator. It is therefore not imbued with the public interest concerns which have given rise to the principles of public accountability, objectivity and balance that now govern broadcasting. No case has been made out for subjecting pay-cable to more extensive legal restrictions than now apply to publishing—namely, those against defamation, sedition and obscenity.

Market provision

The services supplied by pay-cable are in the nature of leisure goods. Pay-cable TV therefore differs from basic services like water, gas or electricity. Because it is a luxury, public provision of pay-cable is not justified.

The market should be used to finance and supply cable services. It has the attraction of giving full scope to individual initiative and experimentation, and can respond quickly and flexibly to the need to adjust to changing demands. These characteristics are the opposite of those of regulation. If they are to be sacrificed, any offsetting gains from regulation must be clearly established.

Maximising efficiency

The objective of cable policy should be to maximise consumer welfare, which can be achieved most effectively by promoting competition. Since the components of consumer welfare are manifold and costly to produce, this objective can be satisfied by measures that allocate resources efficiently. The goal should be to maximise the difference between the value that the consumer places on the quality, variety and quantity of products and the costs of their production.

The data are not always available to enable an efficient solution to be determined with scientific precision. Qualitative evidence can, nonetheless, help to decide whether one arrangement is superior to another. The overriding presumption of cable policy should be that market-supportive interventions which harness competition and the price mechanism provide the surest protection for the consumer.

Cost-effective regulation

The identification of a potential market deficiency does not, *ipso facto,* justify regulation. It must further be demonstrated that the benefits of intervening to correct it exceed the costs of regulation, thereby bringing about a net improvement in consumer welfare. Those costs include both direct expenditures and the 'hidden' losses from diminished productivity, resource misallocation, reduced incentives, delay and the risk of 'capture'. Even when the costs of regulation are justified, they should always be minimised. The most cost-effective methods should be sought to correct market imperfections, and the regulatory process should be so designed as to build in incentives to ensure that controls are cost-effective. Our recommendation to policy-makers is simple: When regulation is proposed, be suspicious and resist it unless the gains are clear and substantial.

Regulatory parity

Maximising consumer welfare should be the goal of government, not only for cable policy but for telecommunications generally. No sector of industry should be unfairly protected from competition, especially by artificial barriers stemming from government licensing. Cable should be seen as one part of the telecommunications market and governed by one integrated policy.

2. Consistency of Government Policy

These principles are broadly consistent with the Thatcher Government's policy towards telecommunications.[1] In the Commons debate on the Hunt Report, the Home Secretary stated that the Government

> 'believe that in many respects private investment and market forces should determine the pace at which and the directions in which there is development [of cable]'.[2]

He went on, however, to endorse the approach and recommendations of Hunt, thereby implicitly contradicting the Government's view.

In the same debate, Mr Kenneth Baker, the Minister for Industry and Information Technology, announced proposals which not only suggested that the Government thinks it knows better than the market which technology should be used in cable,[3] but also introduced significant monopoly elements into its policy. BT and Mercury are to have a monopoly over voice services ('even on cable systems which they have played no part in financing') and exclusive rights to link individual cable systems. Further, the cable *provider* is to be given a licence ranging from 12 to 20 years, depending on the system.[4]

Mr Baker's proposals are clearly inconsistent with a belief in the market and a cost-benefit approach to regulation. Moreover, the method the Government has chosen to implement its policy is unsatisfactory. It appears to be opting for a variant of the American model of public utility regulation by independent commissions. This choice no doubt reflects the predominant assumption in Britain that government will

[1] As stated in the White Paper, *The Future of Telecommunications in Britain,* Cmnd. 8610, 1982.

[2] *Hansard,* Vol. 33, 2 December 1982, col. 419.

[3] *Ibid.,* col. 487. The Minister announced the following 'technology-forcing' standards: mandatory two-way capability, a minimum of 25-30 channels including teletext and DBS service, mandatory audio channels, at least one return video channel, and two-way data channel, TV sets to be compatible with approved cable systems, and every cable system to be compatible with the trunk transmission networks of BT and Mercury.

[4] *Ibid.,* col. 490. This proposal envisages a system of 'split franchises' for both the provider and operator of cable.

faithfully serve the public interest.[1] Economics, however, teaches that the number of policies which rely on faith should be minimised and the inherent inefficiencies of regulation acknowledged—especially when superior and more effective means are available to protect consumers.

The inconsistencies in the Government's policy stem, in our view, from the mix of its belief in the market and its desire to create jobs quickly through cable expansion.[2] To achieve the latter it appears willing, perhaps unwittingly, to sacrifice the interest of consumers. Longer franchises and more elaborate (presumably British) technology are both ways of trying to 'buy' jobs sooner rather than later. But, like all job-creation schemes based on restricting competition and introducing statutory monopoly, it is the consumer who pays. The irony is that so short-sighted a policy will reduce the long-term employment prospects cable expansion can offer Britain. Ultimately, it will benefit neither consumers nor workers as a whole.

3. Outline of a Pro-competition Cable Policy

The policies we propose would minimise the need for regulation and ensure that cable more effectively maximises both consumer welfare and employment. Cable operators are unlikely to enjoy significant market power and methods are readily available to minimise the likelihood that they will acquire it. This argument destroys Hunt's demand for franchising. Moreover, exclusive franchises create artificial barriers to entry which are incompatible with the promotion of competition. There is thus no necessity to control cable operators or providers by franchises.

Since cable operators would not have exclusive franchises,

[1] Professor James Buchanan, Director of the Center for Study of Public Choice at the Virginia Polytechnic Institute, has attributed the neglect in this country of the realities of government regulation to the sway of Benthamite utilitarianism with its 'idealised objectives' for government policy. 'In Britain', Buchanan has written, 'you surely held on longer than most people to the romantic notion that government seeks only to do good . . . and, furthermore, to the hypothesis that government could, in fact, accomplish most of what it set out to do.' ('From Private Preference to Public Philosophy: The Development of Public Choice', in *The Economics of Politics*, IEA Readings No. 18, IEA, London, 1978, pp. 3-4.)

[2] 'The reason we [the Government] want to move quickly is [that] with cabling more jobs will be created.' (Mr Kenneth Baker, *Hansard*, Vol. 33, 2 December 1982, col. 494.)

there would be competition for the same customers in the adjoining areas of different cable companies. 'Poaching' would serve as a market test of the extent to which a cable operator was providing a satisfactory service. 'Overbuild' would be feasible since no cable operator's investment would be legally protected by an exclusive franchise. The Government would no longer face the impossible task of determining the optimal area for a cable system and its associated investment—which Hunt admits is 'a leap into the dark'. The salutary effects of the threat of competition have already been demonstrated in Britain with British Telecom, another alleged natural monopoly; the Government's decision to license a private long-distance telephone company (Mercury) has led to a 35 per cent reduction in BT's inter-city telephone charges.

Break-up BT

Competitive pressures on cable operators can be further enhanced by breaking up the privatised British Telecom into a number of separate local telephone companies. This would increase not only the potential competition faced by cable operators but also that faced by the telephone companies. The proposal would further the Government's policy of liberalising telecommunications by rectifying its principal weakness—namely, that British Telecom is to be de-nationalised as a single company.[1]

Maximise competition between cable and telephone

Local telephone companies should be permitted to offer cable services *provided* a well-established cable company already exists in the area. They would not be allowed to provide a cable service first because of the danger that their superior facilities and sunk costs would form a significant barrier to entry. In return, all cable companies would be encouraged to offer a voice service and to interconnect with one another using long-distance microwave carriers or other means, thus creating an additional telephone network.[2]

[1] Robert Miller, 'Denationalising British Telecom: One Step in the Right Direction', *Journal of Economic Affairs,* Vol. 3, October 1982, pp. 56-8.

[2] A similar proposal has been put forward for the USA by Eli M. Noam: 'Towards An Integrated Communications Market: Overcoming the Local Monopoly of Cable Television', *Federal Communications Law Journal,* Vol. 34, 1982, pp.209-57.

De-regulate broadcast TV

The case for de-regulating broadcast TV should also be given serious consideration. The objective need not go as far as a full-blown market in frequencies; but there should be at least a thorough review of their present allocation. This review should be undertaken by an outside 'auditor' rather than by the Home Office or the IBA, and should pay particular attention to the economic benefits foregone in allocating frequencies by current methods. Consideration should also be given to auctioning existing ITV franchises, to making available more frequencies for broadcast TV, and to allowing subscription TV. The purpose of a policy of de-regulating broadcast TV is not simply to privatise it but to ensure that the costs of the current restrictions are justified and that pay-TV becomes more competitive.

To this end, we advocate that all new broadcasting regulations be subjected to a cost-benefit test to demonstrate that the costs of each do not exceed the estimated benefits.[1]

Granting wayleaves

There is, notwithstanding, a strong argument for new legislation. A government department (such as the Home Office) should be given the power to grant wayleaves or override local authorities' decisions on wayleaves. Moreover, the grounds for a refusal to grant wayleaves should be severely limited to avoid unjustified central regulation. Alternatively, the more radical step of abolishing local authorities' (and central government's) power to grant wayleaves could be considered.[2] The purpose of these measures would be not to control the industry but to prevent local authorities from regulating it in a way that is inconsistent with the objectives outlined above. In addition, cable operators should pay for all wayleaves and be liable for any damage to property or person which results from laying their cable.

[1] As is now done by the Health and Safety Executive. (Health and Safety Commission, *Cost/Benefit Assessment of Health, Safety and Pollution Controls:* Discussion Document, 1982.)

[2] Numerous wayleaves already exist which cable companies could use: for example, those of British Telecom, water and gas pipelines, and sewers. (Guy de Jonquieres, 'Communications gold mine under the streets', *Financial Times,* 18 May 1982.)

Control via competition legislation

If, after all these positive steps have been implemented, there remains evidence of market power among cable operators, the public interest can be safeguarded by existing legislation. The Office of Fair Trading and the Monopolies and Mergers Commission already have the powers to minimise the risk of monopoly. If the legislation is too weak for the task, the solution should be to strengthen it rather than to create a new regulatory agency.

TOPICS FOR DISCUSSION

1. Do you agree with the authors that a market in broadcast TV is feasible?

2. Evaluate the economic merits of Herzel's proposal to auction broadcasting frequencies.

3. 'Cable TV is not a typical natural monopoly requiring public utility status and/or extensive public regulation.' Discuss.

4. Describe and analyse critically the 'local monopoly problem' in the supply of cable TV.

5. 'Cable operators can exploit the consumer only if they enjoy a monopoly in both the geographical *and* the product markets.' Explain and assess.

6. What advantages does pay-TV possess over advertising and the BBC-type licence fee as a means of financing television?

7. 'Commercial pay-TV will tend towards neither low-quality, nor high-quality, nor mass-appeal programmes, but towards the most profitable degree of variation in quality.' Discuss.

8. Describe and evaluate the concept of 'capture' in the so-called 'life-cycle' theory of regulatory behaviour.

9. Discuss the implications for economic welfare and income distribution of permitting cable operators to obtain exclusive rights to national sporting events.

10. What are the drawbacks of awarding exclusive franchises to cable operators?

FURTHER READING

Altman, Wilfred, Thomas, Denis, and Sawers, David, *TV: From Monopoly to Competition — and Back?*, Hobart Paper 15, Institute of Economic Affairs, London, Revised Edition, July 1962.

Caine, Sydney, *Paying for TV?*, Hobart Paper 43, Institute of Economic Affairs, London, 1968.

Coase, Ronald H., *British Broadcasting—A Study in Monopoly*, Longmans Green & Co. for the London School of Economics, London, 1950.

——, 'The Federal Communications Commission', *Journal of Law & Economics*, Vol. 2, October 1959, pp. 1-40.

Howkins, John, *New Technologies, New Policies?*, British Film Institute, London, 1982.

Noll, Roger G., Peck, Merton J. & McGowan, John J., *Economic Aspects of Television Regulation*, The Brookings Institution, Washington DC, 1973.

Owen, Bruce M., Beebe, Jack H. & Manning, Willard G., *Television Economics*, D. C. Heath & Co., Lexington, Mass., 1974.

Williamson, Oliver E., 'Franchise Bidding for Natural Monopolies—in General and with Respect to CATV', *Bell Journal of Economics*, Vol. 7, Spring 1976, pp. 73-104.

Some Recent IEA Publications

Hobart Paper 93

Land and Heritage: The Public Interest in Personal Ownership

BARRY BRACEWELL-MILNES

1982 £3·00

'Barry Bracewell-Milnes has made a valuable contribution to the debate by providing well researched economic—rather than political—arguments for the retention and encouragement of private ownership of land and other heritage assets.'

Country Landowner

'. . . it is good to welcome a carefully-argued case for "the public interest in personal ownership".' *The Field*

Hobart Paper 94

Will China Go 'Capitalist'?

An economic analysis of property rights and institutional change

STEVEN N. S. CHEUNG

1982 £1·50

Professor of Economics at the University of Hong Kong, Cheung draws on his deep understanding of the historical roots of Chinese society and predicts that the country is on 'the road to capitalism'. Because of the current 'open door policies', Cheung contends that China will eventually adopt a structure of property rights similar to Hong Kong and Japan.

Hobart Paper 95

Transport without Politics . . .?

A study of the scope for competitive markets in road, rail and air

JOHN HIBBS

1982 £2·50

'In his far-ranging analysis of the scope for competitive markets in road, rail and air transport, economist John Hibbs goes far beyond the radical ideas for transforming the roads economy, which have been backed by the civil engineering contractors.

The road builders want to see private funds devoted to financing major highways which might otherwise never see the light of day thanks to cutbacks in spending imposed by the government.

The author sets out to dispute the "conventional wisdom", endorsed, he claims, across the political spectrum, that resources in the transport industry are better allocated by administrative decision than by the market'. *Contract Journal*

'Out today is a provoking little pamphlet with the hopeful, if impractical title of . . . Ahead of its time, of course, but this year's rail stike has triggered off much new thinking about the future shape, and cost, of transport in Britain. Some of this book's ideas may be a bit closer than the NUR or ASLEF think'. LEITH McGRANDLE, *Daily Express*

Occasional Paper 63

The Welfare State: For Rich or for Poor?

DAVID G. GREEN

1982 £1·20

'The middle classes often do better out of the Welfare State than the poor for whom it was intended . . . the middle classes are frequently chief beneficiaries of many State welfare handouts, even though these are paid for to a large extent by taxes on the less well-off'. *Daily Telegaph*

'The Welfare State is failing the low-paid people it was designed to help and should be scrapped . . . .Former Labour Councillor Dr David Green claims middle-class people do better out of the system. Dr Green, now a research fellow at an Australian university, says that middle-class people tend to benefit more because they know how to manipulate the system in their own interests.' *Daily Mail*

Occasional Paper 64

The Pleasures and Pains of Modern Capitalism

Thirteenth Wincott Memorial Lecture

GEORGE J. STIGLER

1982 £1·00

'Excessive government regulation hampers the workings of the free market economy. There is nothing, of course, new about this. . . . And it is just what you would expect from this year's Nobel Prize-winner in economics. But Professor Stigler . . . goes one stage further than this.

State regulation of industry, he argues, is there with the full consent of business. And businessmen, despite their public posturings, would soon raise a fuss if the degree of regulation were reduced.' *Financial Weekly*

Occasional Paper 65

How Much Freedom for Universities?

H. S. FERNS *With an Economic Commentary* by JOHN BURTON

1982 £1·50

'State funding of universities, which now accounts for 70 per cent of their income, should be phased out. Instead universities, polytechnics and colleges should become independent companies free to sell whatever course they like for whatever price they can command.

This radical shake-up of British higher education is the only way to rescue academic institutions from the damage now being done to them, according to [this] pamphlet.'

AURIOL STEVENS, *Observer*

‘As one would expect in a report from this quarter, its proposals are designed to be provocative beyond the immediately practical. That is not to say that its analysis is not acute’.

Daily Telegraph, in an Editorial

Research Monograph 37

The Moral Hazard of Social Benefits

A study of the impact of social benefits and income tax on incentives to work

HERMIONE PARKER

1982 £3·00

‘Every politician should read it for the devastating evidence it provides’.

RONALD BUTT, *The Times*

‘Mrs Hermione Parker demonstrates persuasively in a monograph published today by the Institute of Economic Affairs that 5·5m people, or 20 per cent of the workforce, are now close enough to the [unemployment and poverty] traps for their morale and motivation to be affected.’

Financial Times, in an Editorial

4 89 110PT2 3 097 BR 6062